Invisible Student Scientists

How Graduate School Science and Engineering Programs Shortchange Black, Hispanic, and Women Students

Robert Leslie Fisher

University Press of America,® Inc.
Lanham · Boulder · New York · Toronto · Plymouth, UK

4501 Forbes Boulevard
Suite 200
Lanham, Maryland 20706
UPA Acquisitions Department (301) 459-3366

10 Thornbury Road
Plymouth PL6 7PP
United Kingdom

Printed in the United States of America
British Library Cataloging in Publication Information Available

Library of Congress Control Number: 2013951556
ISBN: 978-0-7618-6258-1 (clothbound : alk. paper)
eISBN: 978-0-7618-6262-8

Dedicated to the memory of my father Sidney Solomon Fisher,
a manager who always bet on his employees to do their best.

TABLE OF CONTENTS

List of Tables

FOREWORD

CHRISTOPHER SMITH, DIRECTOR OF RESEARCH AND PROGRAM EVALUATION, NATIONAL ACTION COUNCIL FOR MINORITIES IN ENGINEERING, INC. (NACME)

The future of science, technology, engineering, and mathematics (STEM) in the United States depends on our nation's ability to engage diverse populations at all stages of the educational pipeline. While underrepresented minority groups (URMs—African American, Latino, and American Indian) make up 29 percent of the U.S. population (and are expected to grow to 43 percent by 2050), they only make up 14 percent of the total STEM workforce (NACME, 2011). Underrepresented groups must be engaged with STEM early in their academic careers to counter this trend. The National Action Council for Minorities in Engineering, Inc. (NACME) works with partners from middle school through workforce entry to increase the proportion of URMs in STEM education and careers. NACME, led by President and Chief Executive Officer Dr. Irving Pressley McPhail, is widely known as one of the nation's largest private sources of scholarships for African American, American Indian, and Latino women and men in engineering. To date, NACME has provided $124 million in scholarships and support to more than 24,000 underrepresented minority-engineering students at 160 colleges and universities across the country. However, we've learned through experience that scholarships alone do not suffice in helping more URMs complete their STEM postsecondary degrees.

For those URM students who enter college majoring in a STEM discipline, the dropout rates are alarming. The Higher Education Research Institute at UCLA tracked a cohort of students who enrolled in college in Fall 2004 and initially majored in a STEM discipline, and the five-year degree completion rate was 22.1 percent for Latinos, 18.4 percent for African Americans, and 18.8 percent for American Indian STEM majors. While there are numerous factors that contribute to these low completion rates, one that is often overlooked is the campus environment. Our experience working with Partner Institutions has shown us the importance of creating an optimal learning environment for URMs to persist and succeed in attaining their STEM degrees. In Colleges of Engineering across the country, Minority Engineering Programs (MEPs) help to establish this environment by helping underrepresented groups become integrated with the college. In addition, institutional leadership must be committed to recruiting and admitting promising students from underserved communities and two-year community colleges. Admissions policies must reflect a holistic approach that goes beyond standardized test scores in evaluating student potential for success. Summer bridge programs should be offered to enrich intellectual exchange and socialize students for participating in the life of the university and the STEM community. Finally, a campus community and institutional support structure is needed for faculty, students, and administrators, which helps increase student engagement and grows the capability to graduate more URMs in engineering who are prepared for entry to graduate school and the workforce.

The lessons we've learned with our Partner Institutions can be applied to the graduate level as well. Robert Leslie Fisher's research provides important insight into the challenges that women and URMs face when attending graduate school. Oftentimes, the support structures that are in place for these groups in their undergraduate programs are not present for them in graduate school. Fisher is concerned that research universities are not actively recruiting African ancestry students and other URMs. He also believes that an abundance of men in these fields—both American and foreign-born—maintain biased attitudes towards women in the sciences that universities have not done enough to counteract. Fisher argues that this combination of issues is likely to leave the URMs and women feeling isolated, lacking a sense of self-efficacy, less inclined to complete their doctoral studies, and more likely to seek less prestigious and less remunerative teaching posts if they do graduate. Much like MEPs foster a welcoming environment for URMs in college, programs must be developed by universities to support women and URMs who are pursuing an advanced degree. NACME manages The Alfred P. Sloan Foundation's Minority Ph.D. Program, which offers substantial scholarship support to underrepresented minority students who are pursuing their doctoral work in engineering, natural science and mathematics. We also manage the Sloan Indigenous Graduate Partnership (SIGP), which allows eligible American Indian masters and doctoral candidates at selected universities to apply for scholarships designed to facilitate graduate study in mathematics, natural sciences, and engineering through approved faculty and departments. The Sloan Foundation also recognizes the need for additional supports for underrepresented groups, and selects a Program Director at each university who helps to promote the success of URMs. The Program

Director ensures that each student is paired up with a mentor. These mentors lead by example, guide students through the requirements of their programs, and help students develop the complementary skills that are needed for success. Sloan Scholars are also invited to the annual Institute on Teaching and Mentoring, which is the nation's largest gathering of minority Ph.D. students who are seeking careers in academia. These types of support structures are needed to help develop new talent in the field. Too often, women and minorities are made to feel like outsiders in the academic environment, so universities, policy makers, and funding organizations must work together to solve this issue. Mr. Fisher's research leads him to call for a sea change in educational policy and practice. This is necessary not only to create an equal chance for all groups to succeed in STEM, but also to help advance U.S. competitiveness in these fields.

PREFACE

It is my custom to do a Preface for my book length monographs and I have followed that custom here. I began the process of preparing it by looking over the Prefaces written for earlier works. When I reread the Preface for *Crippled at the Starting Gate* (hereafter *CATSG*), the previous monograph in this two-part series, I was both elated and saddened. I was elated because its message seemed so relevant some years later and I thought might not have to alter it much for it to be useful in this book. And yet I was also saddened that its message was still relevant.

Of course things are not identical. Everyone from President Obama on down who worries publicly about our nation's military preparedness and our economic strength agrees that we need to strengthen science education as a key to keeping our nation strong and prosperous in the future. That is a good development.

And yet in other ways too little has happened and some developments are truly alarming. One such alarming development is the growth in efforts to recruit foreign graduate students, especially from countries such as China, India, Korea, and other economic rivals of the United States. The motivation is money, but it is a development with quite dire implications for the future. Who will do the innovative product development in the future if we are training students who are going back to China and India to help their high tech companies? Who will do the military research at the frontiers if we are training people from potential military adversaries to help their defense industry? How is it good economics to stuff our graduate science and engineering programs with foreign students even if they pay full tuition?

Student tuition in the sciences covers only a fraction of the true cost of educating students, unlike programs in business or law schools that are cash cows for universities.

Recruiting hordes of foreign graduate students in the science and engineering programs is bad enough. But it is accompanied by indifference to remedying the low representation of African American and Hispanic American graduate students in the STEM fields.

How serious is the low representation of African American and Hispanic American graduate students in the STEM fields? So few show up in these graduate programs that it requires great care—and no little additional expense—to find enough of these minority STEM graduate students to say anything much about their perceptions of their graduate school experience and how it can be made better. I have to ruefully admit that for *CATSG* I simply had too few African Americans and Hispanic Americans in a sample of nearly 1,100 respondents to report much about these science graduate students. I considered it a major drawback of my prior study that I was determined to remedy it in a future volume.

Invisible Student Scientists is that future volume. It is a very different book in some ways from CATSG. Though it was not my intention when I began this project, I had to do a special sampling of 60 major research oriented graduate schools to find the 300 respondents for this study that are the major focus of the book. Even so, my respondents are not mostly native-born African Americans and Hispanic Americans. A large group of them are foreign born individuals.

Including foreign-born students in the study was a controversial decision. How externally valid is a sample of students that is so heavily foreign born? I cannot answer that question with certainty. But I can say that this study would have been impossible to do without including them. I believe the experiences of these foreign-born, and often female, students are worth reporting, providing useful guidance for academic officials and professors who will be educating STEM students who are nonwhite and increasingly female instead of the white male students that the schools have generally educated in the past.

Educating African American and Hispanic scientists is not the right thing to do from just a justice standpoint. It is also the right thing to do from the standpoint of our national survival. We must have a pool of scientists and engineers to do the research and product development work that will keep us militarily secure and economically prosperous. That requires two things of us as a society. On one hand, we must invest in our universities to equip them to train people willing to sacrifice years of their working lives, an expensive proposition, but an expense we must bear. On the other hand, we must have a pool of people willing to undergo this rigorous training over many years. We already know what will happen if we fail to do this. A very long time ago, a famous British historian, Edward Gibbon, M.P., said it well in his observations about the ancient Athenians:

> In the end, more than freedom, they wanted security. They wanted a comfortable life, and they lost it all - security, comfort, and freedom. When the Athenians finally wanted not to give to society but for society to give to them, when the freedom they wished for most was freedom from responsibility, then Athens ceased to be free and was never free again.

For decades, white male students have filled the ranks of graduate students in science and engineering. But the white males are not coming in the numbers that are required by our postmodern economy. White male students with promise in mathematics and science increasingly are heading to law, or business school to get degrees that they can use to command high salaries. And they can get business degrees in two years and law degrees in three years. In contrast, a typical doctoral student in a STEM field can look forward to seven years of post college education followed by years of postdoctoral fellowships and low paying university junior positions. It is no wonder the schools are not graduating enough future scientists. Even as the need for statisticians, information scientists, electronics engineers, and chemical engineers grows, we are dependent on foreign scientists to fill industry and academic jobs because enough qualified Americans are not available.

As in CATSG, my intent in this book is to assist schools in recruiting and educating more students who are women and/or nonwhite. And as I pointed out in closing the Preface of CATSG, "The hour is late in the struggle for American competitiveness in the world of ideas". We in the United States have a long way to go to realize the hopes of Vannevar Bush and other farsighted scientists and educators that we will achieve a leading position in science among the world's nations by drawing on the talents of all our people, regardless of gender, race, or ethnic origin. My hope is that it is not too late and this book plays a positive role in bringing more of our best women and minorities into science and engineering and rewarding them with satisfying careers.

Delmar, New York
April 2013

ACKNOWLEDGEMENTS

As is the case in all my books, I have benefited from the assistance of numerous people. It is my pleasure to acknowledge my gratitude here.

As in my previous monograph on the subject of graduate schools' encouragement of students to successfully complete their studies, I must first acknowledge the assistance of Dr. Edward Halteman, a mathematical statistician and survey research consultant whom I hired to do the data collection, for his brilliant work in obtaining a large group of African ancestry and Hispanic heritage graduate student respondents. For data analysis, I relied on Dr. Chris Galvan, who ably carried out my wishes regarding the formatting of the tables and made many useful observations about what tabular data would be useful to present. Joanne Ruppel provided her usual skillful work in taking my manuscript and putting it into acceptable shape for publication.

I also want to thank the following for their comments on the book—Dr. Wendy Williams, a Professor of Human Development at Cornell University and the National Action Council on Minorities in Engineering. Despite their best efforts, I am sure the stock of errors that remain is large and I accept full responsibility for them.

I also hereby acknowledge the contribution of the nearly 1,400 student respondents who patiently answered my questions and whose stories are the content of this book.

Finally, I thank my lovely bride Shirley Anderson because she creates the wonderful environment that permits me to write. In everything she does, she embodies the ancient Jewish saying that "A wise woman buildeth the home."

INTRODUCTION

The United States has an urgent need to fund graduate education in the sciences and education. Graduate education in sciences and engineering produces the scientists and engineers who will keep the United States a leader in critical technologies that are essential to the nation's prosperity and military security, e.g., discovering the drugs that will cure currently incurable diseases and inventing the artificial intelligence that will allow currently dangerous tasks to be done by machines rather than by humans.

Yet American students are spurning graduate school in science and engineering programs even as the need for people with advanced skills grows. The nation's problem of producing too few scientists and engineers is worsening rather than improving because the graduate schools are actively recruiting abroad in China and India to fill their empty seats. The graduate schools see this as not only a way to fill empty seats but even more important a way of gaining high tuition paying students. It is a fool's bargain, however. Massive recruiting from countries such as China and India trains the scientists and engineers who will power the economies of the countries from which these students have come—in the case of India and China already formidable competitors in many spheres to American companies. Massive recruiting from such countries may also be producing the engineers and scientists who will work for the defense industries of potential adversaries in a dangerous and volatile world.

Yet another problem with such massive recruiting of graduate science students from countries such as China and India is that a high proportion of these students—perhaps as much as thirty percent—arrive with attitudes that women find frankly sexist and racist, e.g., beliefs that men make better professors and scientists. They *compound* the problem of an environment that is unwelcoming to women—including foreign women—and students of African ancestry and Hispanic heritage. The racism and sexism problem is especially pronounced among foreign men students and is more severe in the hard sciences and certain engineering specialties (interestingly, in the soft sciences, the foreign students often are as liberal as Americans in their attitudes toward women as scientists and professors).

It is important to emphasize that the foreign students did not create a graduate school environment that is unwelcoming to women and minorities. It was here long before the schools began recruiting foreign students. The fact is, the United States has a two-class graduate school system in the sciences and engineering. We see its consequences wherever we look: white male investigators win more research grants from the National Institute of Health than African ancestry men scientists with the same credentials (Ginther et al., 2011). Compared to women, more white men enjoy private foundation funding, a preferred form of funding because it typically requires less time spent on grant application preparation and administration (Xie and Shauman, 2003; also Fisher, 2007). Although most science today, as Mary Frank Fox (1995) points out, is collaborative work women have fewer collaborators for research than men (Bozeman and Corley, 2003); Bozeman and Lee, 2004)

Thanks to their contacts, their memberships in discussion groups doing the cutting edge research and other intangible resources that comparatively fewer women and African ancestry or Hispanic heritage colleagues possess, the white male students who finish their graduate school studies successfully go on to fine careers in academia. White men publish far more than women or men of other ethnic and racial groups—the well-known gender gap documented in dozens of studies over the past fifty years (see e.g., Fisher, 2007; Xie and Shauman, 2003; and Cole and Zuckerman, 1984). And not surprisingly, in a world which rewards greater scientific output, regardless of quality as measured by citations (Valian, 1998) white men dominate the faculty of the university science and engineering departments—often representing over ninety-five percent of the senior faculty and rarely less than ninety percent.

That this two-class system exists at all is reprehensible. It offends basic American values of fair play and equal opportunity for all enshrined in the Declaration of independence and the Constitution of the United States, our nation's founding documents. That this two-class system exists in science is perhaps even more appalling since science prides itself on being a merit based system. If this is not sufficient reason for reform now, consider the fact that this two-class education system threatens our nation's competitiveness in a global economy. We need to be producing the highly trained scientists and engineers who will be inventing the machines of the twenty-first century and making the scientific discoveries that will permit us to innovate our way out of challenges such as air pollution, water pollution, desertification, rising seas, global warming and so on. If we do not rise to this challenge, we will need to continue importing scientific and engineering talent—as some of our nation's business and political leaders have urged.

A report issued in May 2012 by Partnership for a New American Economy—a group sponsored by Michael Bloomberg, the billionaire Mayor of New York City—called for

> fast tracking permanent visas for immigrants who have advanced degrees in such areas as science, technology, engineering and mathematics and visas for foreign students who are studying for their advanced degrees in American universities. (Associated Press; see, for example, "Bloomberg says tap world for talent," Albany Times-Union, May 23, 2012).

Yet as a result of obsolete or merely wrongheaded policies right now we are producing too many college graduates with too few skills that employers want (see, for example, the Associated Press article of April 23, 2012); training foreign competitors in our nation's premier science and engineering graduate programs; and pushing women and minorities out of science and engineering from high school right up to the time the comparative few who achieve their doctorates embark on their careers as scientists and engineers.

This sorry state of affairs did not happen overnight. White men's dominance of American science is the result of a confluence of historical forces—graduate schools as traditionally male bastions that initially excluded women and minorities; disciplines that sought professional status at a time when women were disenfranchised and therefore sought to limit the number of women in the discipline; and intellectual theories that suggest that white men's advantage is somehow the result of innate advantages of white men.

Are white men superior scientists, endowed with the "right stuff" as Pinker (2005) suggests? Or is the game rigged?

Despite Pinker's spirited effort to demonstrate that men have the right combination of intellect and other traits that are needed to do science at the frontiers, many other psychologists do not share his views. Psychologists have shown that women do better in school than men (Halpern, 2007), with higher numbers of women than men receiving college degrees every year since 1982 (p.123). Hyde et al. (2008) showed that women have equal aptitude in mathematics.

Pinker would retort that men dominate at the very top of the IQ measure, the top one percent. True, but as Halpern points out, "Women are obtaining 50% of the MD degrees from medical schools, almost 75% of he VMDs from veterinary schools and 44% of the Ph.D.s in biology and life sciences." Clearly, Halpern is correct when she concludes, "women are succeeding in some sciences and thus must have the cognitive ability to learn and succeed in science" (p. 122).

If white men's innate traits are not the reason for their overwhelming dominance of the science and engineering departments, how is the game rigged?

Perhaps the game is rigged through bias in research grants awarded by government. Ginther et al. (2011) suggested this possibility in a review of grants from NIH, one of the main sources of federal research funding. NIH's director promised to investigate the agency's procedures to see if bias was present in their grant review procedures and this bias was accounting for the discrepancy in grant awards between white and African ancestry investigators. I am betting that when NIH completes their review they will announce that they found nothing improper. Will this be because of a cover up? I do not think so.

The reason NIH will find nothing wrong in their procedures, i.e., no deliberate bias in favor of white male investigators is that while the game is rigged the problem does not lie in NIH review procedures. Rather, problem lies elsewhere and is not amenable to action from the National Institutes of Health. Only Congress can address the gender (and race) gap in science.

Let me return for a moment to my example of the NIH grants. An economist at the University of Kansas, Donna Gintner, and several colleagues of hers investigated the award of grants by the NIH. On the basis of their study, they published a paper in *Science* that alleged, "From 2000-2006, black grant applicants were significantly less likely to receive research funding than white applicants."

I maintain that the reasons for less federal grant support for work of black or Hispanic or women scientists are not discrimination in the award of funds on the basis of race or gender. Rather, the reasons lie further back in the training of white male students versus other graduate students.

Is it differences in the coursework white men students take compared to other graduate students? Not only is that not the case it is not even plausible as an explanation. All scientists, regardless of gender or ethnic background will take the same courses as their white male colleagues to obtain their degrees as physicists, chemists, or engineers.

Yet the educational experience of the white males is qualitatively superior to that of other component populations of graduate students. Quite simply, in subtle but important ways, the white male graduates in physics, chemistry, engineering, and other sciences have obtained more valuable preparation for their subsequent careers than have their African ancestry, Hispanic heritage, or female colleagues (See Etzkowitz et al., 2000; Fisher, 2010).

The totality of subtle advantages white men enjoy has been beautifully captured in the memorable words of Jonathan R. Cole (1981, quoted in Fox, 1992) "although women have moved into science they are not *of* the community of science." He further noted that women more often than men remain "outside the heated discussions, inner cadres, and social networks in which scientific ideas are aired, exchanged, and evaluated." What Cole has pointedly said of women relative to men applies as well to Hispanic heritage and African ancestry students, and even to male Asian students in comparison to white male graduates—though perhaps to a lesser degree. And as Etzkowitz et al. point out, the isolation of women and minority students begins while they are still in graduate school—a point I will develop in more depth in Chapter 1.

If the literature is so compelling in its documentation of the superiority of the informal education white male students receive in graduate school compared to others, why the need for the present study?

The prime reason for the present study is that we have limited data about the problems of African ancestry students and Hispanic heritage students in STEM discipline graduate programs. Much of what I presented above is anecdotal (e.g. Leonardi, Etzkowitz), qualitative, and merely describes how white male students in graduate school compare to women students, most of whom are also white. The present study I hope will fill some of the gap in our knowledge in this area.

In other words, when I began this research I *expected* to find that African ancestry and Hispanic heritage students fare poorly compared to white male students but I did not have the evidence in hand. This desire to document the problem was one of the impetuses driving the research reported in this book.

The question I am addressing has been recognized as important for a long time. In 1998, for example, a team of educators and economists (Leslie et al., 1998) remarked, "Not only does the lack of parity (between white males and women, and/or between white males and males of other ethnic groups) raise obvious social concerns, it has consequences for the economy." In regard to the economic effects they pointed out,

> The quality of the science and engineering labor force is at issue, for even though the supply of scientists and engineers appears, at least for the present, to be adequate to national needs, larger supply pools translate into higher quality scientific workers, thereby increasing marginal labor productivity and, in turn, the productivity of the overall economy.

In concrete terms,

> All else being equal, we would expect that the more women and minorities enter these (S and E) fields, which are known to be powerful contributors to national productivity, the larger will be the national stock of scientists and engineers, the greater will be the quality of that stock, and the greater will be the productivity of the nation's labor force (Leslie et al., 1998, p. 240).

The present work raises serious questions about the actual contribution scientists and engineers who are not white males are allowed to make in our economy. In general, I can say that regrettably, the African ancestry and Hispanic heritage students fare more poorly versus white male students than do women in the areas I examined.

Before the reader shuts this book and goes off to denounce the racism and sexism of the graduate schools, I need to provide a word of caution. The results reported in this study come from two separate surveys with populations that are not strictly comparable. White students, most women, and Asian students were studied in an earlier survey carried out in the New York City and suburban New Jersey region and the Research Triangle area of North Carolina after the short research instrument was piloted in the Boulder, Colorado area. However, the vast majority of African ancestry and Hispanic heritage students were surveyed in a nationwide survey almost two years later which was done over the internet and included students from sixty research oriented graduate schools.

With minor differences, the research instrument was the same in both cases and I am comfortable in making comparisons of white and other students. However, these comparisons suggest the depth of the problem we face in overcoming the shortage of scientists and engineers with advanced credentials by harnessing the talents of African ancestry, Hispanic heritage, and women students with an interest in and aptitude for science.

While it is impossible in a work of this sort to lay out a comprehensive list of all the steps we need to take to improve the conditions of minority STEM discipline students on campus, I can say that it is naïve to assume all we need to do to increase the numbers of these African ancestry and other students is to install more ladies' rooms in graduate student dormitories for women students, give everyone off on Martin Luther King's birthday, and remind the students that though Genoese by birth and nationality (Genoa at the time was an independent state on the Italian peninsula) Christopher Columbus was in the employ of Spain when he discovered the New World.

Much more likely to work and to be needed is, first, creating more senior professorships that can be filled by women and minorities in the sciences and engineering who can act as role models and mentors for the women, African ancestry and Hispanic heritage students. This will be very expensive but I maintain it is needed to inspire students of these backgrounds to "stick to the thing" as I. I. Rabi, the Nobel Prize winning physicist might have stated it. Graduate schooling is arduous; the road to the doctorate is long (about seven years typically), perhaps two years on average longer than it was when Vannevar Bush (1945) urged the country to welcome more scientists from the ranks of women and minorities.

Second, the schools will need to provide either coursework or study materials describing the contributions of women, African ancestry and Hispanic heritage scientists and engineers that all students will be expected to become familiar with before they can graduate. S and E students need to be aware of these contributions because the environment in which they work is overwhelmingly white and male and tacitly supports the notion that not only is this the way it is but more importantly this is the way it should be.

Third, the schools will need to provide the professors with material on how to avoid behavior that appears sexist or racist and issue clear policy statements that demonstrate that the school is a community of scholars and all who wish to contribute to the advance of knowledge are welcome within its confines.

These suggestions will not solve the problem of too few women and minorities in the top ranks of science and engineering. However, they will create a climate in which a solution becomes possible. And that is essential at a time when demand for highly trained specialists is likely to be rising, in an era when we must compete with foreign countries such as China and India for scientific and engineering supremacy, we have a shortage of well trained scientists and engineers that we are constantly trying to meet by recruiting abroad.

I first broached many of my suggestions about what is needed in my earlier work, *Crippled at the Starting Gate*. However, while *CATSG* made what I believe was a compelling case for some changes in policies and practices to rectify the problem of too few women in science, it lacked a compelling case for changes that might be needed to encourage more scientists and engineers of African ancestry and Hispanic heritage to obtain doctoral degrees in the sciences and engineering or the STEM disciplines generally. Quite simply my sample of over one thousand graduate students had just a handful of African ancestry and Hispanic heritage respondents.

Why was I unable to find more respondents of these two backgrounds? Two possibilities seemed plausible: perhaps, I designed a poor sample to achieve adequate representation of African ancestry and Hispanic heritage students. I rejected that idea because my sampling was done in areas where large numbers of such students were likely to be: the New York City and New Jersey area and the Research Triangle of North Carolina.

The second possibility was that African ancestry students and Hispanic heritage students simply are not attracted to graduate school in the sciences and engineering. Of course, that raises interesting questions of just why that might be the case.

Although I could not answer that question with the data I collected—instead concentrating on the negative aspects of graduate school training of women scientists—I could not forget my astonishment at finding so few African ancestry and Hispanic heritage graduate students. In a real sense, it was the impetus for undertaking this research study.

Of course it is more than simply demographics that cause me to be concerned about the tiny numbers of graduate students in the STEM fields who claim African or Hispanic heritage. It is also a social justice question. Science is proud to claim that it is a merit-based field. And officially, the welcome mat has been out in science for women and minorities since the end of World War 2 when Vannevar Bush in his manifesto *Science: The Endless Frontier* wrote, "We think it very important . . . there be no ceilings, other than ability itself, to intellectual ambition. We think it very important . . . every boy and girl shall know, if he shows . . . he has what it takes, the sky is the limit."

Are the minuscule numbers of African ancestry and Hispanic ancestry graduate students in the STEM disciplines a result of their not having the necessary ability? Or is it the result of a confluence of historical forces and policies that keep qualified people from these groups out of graduate schools? The answer is not really known, and in fact may be unknowable.

We do know that women were kept out of graduate schools in the sciences until the twentieth century. Furthermore, many white schools openly discriminated against African ancestry students although several colleges, even medical schools, beginning in the late nineteenth century did exist to provide education for a "black elite."

This is changing—today, women are flocking to engineering school and majoring in the sciences in college and they are doing this in increasing numbers in various fields (and, in fact, they now outnumber men in undergraduate colleges, though not in the hard sciences) but women are still notably absent from the higher echelons of the hard science disciplines. And in no field including such "soft" sciences as psychology, are women more than about ten percent of the total senior faculty.

It is fine that women are making progress in their efforts to advance in America's professions. However, the focus of the present monograph is African ancestry and Hispanic heritage students. What can we say about their post secondary schooling?

The Higher Education Research Institute at UCLA tracked percentages of college freshmen intending to major in science and engineering fields from 1971 through 2000. They found that in 1971 6.2 percent of freshmen intending to major in one of the S & E fields were African ancestry students. In 2000, the proportion of all freshmen intending to major in an S & E field who were African ancestry had climbed to 11.5 percent, almost doubling over that thirty-year interval. More revealing perhaps is the fields in which African American students were intending to major. In 1971, African American freshmen were ten percent of the total number of freshmen choosing computer sciences. This rose to 15.7 percent in 2000. In 1971, African ancestry students were 4.2 percent of the freshmen intending to major in biological sciences but in 2000 they represented 11.3 percent, nearly triple the proportion in 1971.

If interest in science and engineering reached an all time high in 2000 among African ancestry college freshmen, why are there so few graduate students of this racial background? I do not have the answer to that question.

A more pressing question, however, is what can be done to change this low enrollment of African ancestry and Hispanic ancestry individuals in graduate school S & E programs? I assert that we must begin addressing this issue by first identifying what policies etc. and practices of graduate schools are the strongest incubi for African ancestry and Hispanic heritage students as regards these programs. In my previous monograph *Crippled at the Starting Gate* I pointed to teaching assistantships as a serious obstacle to enhancing the numbers of women STEM disciplinarians at the doctoral level. Teaching assistantships, I found, were more likely to be given to women than men albeit the difference in proportions was not great. But more importantly, these assistantships, especially the burdensome undergraduate teaching responsibilities they carried, interfered with the students' mastery of the material they needed to learn in order to become proficient in their chosen discipline. Do Hispanic and African ancestry students also complain about these problems? This was one of the main questions of interest in this study.

Even if the data in my study sustain my conviction that teaching assistantships are not a good way to finance graduate education, we must not jump to conclusions.

One reason we must be cautious about reaching conclusions and making policy changes is that I followed different sampling strategies in gathering the data for this volume versus the first volume. I will discuss in greater depth the difficulties encountered in planning and executing this study in a later chapter on research procedures. Here, I will simply point out, whereas the white and Asian students were from three metropolitan localities, the African and Hispanic heritage students came from sixty graduate schools across the country.

How should the results of this study be treated? One critic derided as "an act of desperation" my decision to compare the white and Asian students on one hand with the African ancestry and Hispanic heritage students on the other. This critic preferred I just throw away the data and tell the world a tale of woe about the difficulties of studying this subject. I reject that position just as firmly as I reject the idea these data are the final word on the subject of this study. My position is that while we should be cautious in relying on this study alone to influence changes in educational practices and policies, we should be planning individual school studies to try to replicate these results. And if those studies broadly support the results found here, we should act decisively on them.

In this research I focus on perceptions of graduate students of African ancestry and Hispanic heritage. Just as with women, who were the focus of my prior monograph, I take the position that it is policies and practices of the graduate schools that account for our nation producing too few scientists and engineers with advanced training with Hispanic heritage and African ancestry.

Because my focus is on how various policies and practices in the graduate schools affect African ancestry and Hispanic heritage students, I do not provide a nuanced picture of the issue. Part of the problem is simply that unlike with women we have much less information about African ancestry and Hispanic heritage students. Women's problems with science and engineering have been intensively studied for decades. Not only is there a rich sociological literature going back perhaps as far as the 1960s there is also a psychological research tradition. Economists have also made contributions, as have historians.

In the case of Hispanic heritage and African ancestry graduate students almost nothing is known. I believe this survey is unique in its focus on these students. Not only is it unique, I know of no planned studies that will cover the ground it has trod. As with any first study this monograph will eventually come to be regarded as obsolete in many details and perhaps even plain wrong. However, we have to start somewhere and I believe there is value in trying to replicate with African ancestry and Hispanic heritage graduate students a study that illuminated the situation of women graduate students.

Yet, while I think my study has value I also want to acknowledge that it is incomplete in its analysis of the problems of graduate schools. That is, while what I say is correct in my opinion, it does not even pretend to tell the whole story. For example, I might start with my basic assumption: the reason I find so few African ancestry and Hispanic heritage students in graduate school in the sciences is because of a variety of policies and practices that discourage them from attending. Beyond what I know about the graduate schools' attitudes toward women I know of no studies that have ever looked at this issue. I think this study adduces enough evidence to show my hypotheses have some merit—graduate schools' misguided policies and practices contribute to low enrollment of African ancestry and Hispanic heritage students. However, even advocacy groups see some reason to question that it is graduate school policies and practices that are the main culprit.

In the case of African Americans, Hispanic students (also referred to as Latino) and American Indian students, the National Action Council on Minorities in Engineering (NACME) a research organization that has been studying the issue for some thirty five years, makes clear the problems at the graduate school level is a minor aspect of the total problem of low numbers of individuals of these ethnic and racial groups in the STEM disciplines. They reported "Fewer than one-in-three African American students were retained to a bachelor's degree within six years of starting their degree programs" (p. 4).

As NACME clearly show, until this huge attrition at the under graduate level is substantially ameliorated we will not have a large pool of African Americans for graduate training. Still, if my study perhaps fosters the erroneous impression that merely by making some needed changes in graduate school practices and policies we somehow will see a tidal wave of African ancestry and Hispanic graduate students, I still believe it is a necessary contribution to the literature if it leads to a better educational experiences for these students. It behooves us to provide the best possible educational experience we can for the hardy few African Americans and Hispanic heritage students who manage to surmount the hurdles of their prior socialization in order to study science and engineering at the graduate level. We have a shortage of well trained scientists and engineers at the graduate level and we should not aggravate that problem with policies and practices that drive qualified people—of any ethnic or gender background—out of the STEM fields.

Chapter One
Theoretical Orientation and Study Plan

This is the second monograph I have authored on the subject of education at the graduate school level in the sciences and engineering. The first, Crippled at the Starting Gate (2010), argued that the gender gap in science has its roots in the graduate school experience of students of diverse backgrounds. Graduate school policies and practices, I maintained, (1) drove more women from science and engineering than men and (2) these policies and practices gave white men a significant advantage in their careers over others when it came to publishing papers. By offering more resources to white men than others, white men could do more papers—though not necessarily better papers. Some research (Valian, 1998) shows women's publications are better as measured by citations per paper. Since quantity of publications weighs more heavily in tenure and promotion decisions than quality, white men would have an advantage in their careers in such areas as tenure, promotions and academic honors.

If the schools are simply weeding out people who are inferior and will not contribute meaningfully to science and engineering why should we condemn the universities for this? Should we in fact not applaud their steps to remove the untalented?

Weeding out the unfit and the uninterested is a necessary role that the universities play. However, we must always keep in mind that scientists are made, not born. It is just as important to avoid pushing out talented people because they were not given a fair chance to demonstrate their skills and aptitude, as it is to separate from the school those who show no talent or motivation to succeed.

It is even more urgent now that we not have a perverse weed out system that pushes out more women than men, or more nonwhite than white students because the need for scientists with advanced credentials is larger than in the past and growing.

We should not lay all the blame for a shortage of scientists and engineers with advanced credentials at the feet of a perverse weed out system. The causes of women dropping out from graduate school and their careers are many and complex. And they differ somewhat from the causes for the low numbers of minorities in graduate programs. However, I maintain in this study that some of those reasons for women dropping out and African ancestry and Hispanic heritage students not coming (or leaving if they do) are related to policies and practices of the graduate schools themselves.

This is an opportune time to examine graduate school policies and practices. Women, and minorities are graduating in increasing numbers from undergraduate educational institutions with bachelor's degrees in sciences, but the fact remains they are not represented in graduate schools in the sciences and engineering in numbers anywhere near their proportion in the population or even in proportion to their numbers within the population of people graduating from undergraduate institutions in the sciences and engineering.

Literature Review: Nature Versus Nurture

I am an unrepentant "extreme nurturist" as Steven Pinker defines the term (see Pinker and Spelke, 2005) in regard to the gender gap in science. However, while I favor a sociological explanation and rely on the social science literature—in particular the theoretical ideas of Cole and Singer (1992)—I think it is merely parochial to ignore the fact psychologists have also contributed to our understanding of the gender gap in science.

Cole and Singer emphasize their intellectual debt to the psychologist Albert Bandura (1977). In the intervening decades since Bandura's seminal work, other psychologists (who in varying degrees are nurturists as regards the gender gap in science) have added to our understanding of the gender gap as well.

Halpern points out that women are different in some respects from men in their interests and aptitudes. She cites a study by Lippa (1998) showing a difference in the "relative strength of their interest on the 'people versus things' dimension of individual differences scale (Lippa, 1998), with more women toward the 'people' end and more men toward the 'things' or mechanical end."

Eccles (2007) has remarked, however, that this should not automatically lead women to disdain the hard sciences since working in these fields can "fulfill humanistic and helping values." In short, women's differences in interests from men's need not be a basis for women's rejecting a science or engineering career. Proper counseling can influence more women to seek careers in those disciplines.

While psychologists have made important contributions to the nurturist perspective through their research and theoretical insights, a few psychologists, (who I refer to as "naturists") have hunted for evidence that psychological differences between women and men somehow explain the gender gap in science. This quest has ancient origins: from its very beginnings in the nineteenth century some members of the discipline have engaged in this hunt for proof that women are less intelligent than men. Ehrenreich and English (1973) state, "a well known 1848 obstetrical text . . . stated 'She [Woman] has a head almost too small for intellect but just big enough for love'" (p. 29). (Alas, the arguments the naturists advance nowadays to explain the gender gap in science and engineering are more sophisticated than those of their nineteenth century predecessors but nowhere near as amusing).

Two recent papers by psychologists (Pinker, 2005 and Williams and Ceci, 2012) are germane in this regard. I want to first discuss these papers, especially Williams and Ceci, before I present my theoretical perspective. After that, I will discuss how I plan to test the Cole and Singer theory to see if I can confirm or disconfirm its main hypothesis that prior socialization plays a major part in decisions to drop out of the science pipeline.

Pinker looked at the gender gap in science and simply could not imagine that the huge gender disparity in men and women at the top of the academic world was the result of social arrangements alone. He believes that those of us (myself included) who adhere to this notion that social arrangements alone have caused this disparity are "extreme nurturists." He is in favor of a more balanced approach that acknowledges that differences in psychological traits of men and women play some role in this situation as well. Pinker cites the fact that women are not numerous in the very top group of people with mathematical aptitude and that women are not as strong as men in spatial reasoning; strong mathematical aptitude and spatial reasoning, It is these latter psychological traits, Pinker insists, that characterize scientists who are working at the frontiers (Pinker, 2005). Women also are less likely to be risk takers, less likely to be eager to work extremely long hours, and less likely to like making lots of money than men according to Pinker.

I will refrain from critiquing Pinker's views here as I have already addressed this issue elsewhere (see Fisher, 2007, and, for a full statement of his point of view, see Pinker and Spelke, 2005). Instead, I will turn to a more recent naturist perspective (Williams and Ceci).

Williams and Ceci hypothesized that women are more interested in having children than men and willingly leave the workplace because they want to be mothers more than they want careers in science. They assert, "The effect of children on women's academic careers is so remarkable that it eclipses other factors in contributing to women's under representation in academic science."

As evidence, Williams and Ceci point to anecdotal material and extensive research by other psychological and social scientific investigators. For instance, they tell the story of "Jennifer," an "extremely talented" woman. "As a 33-year-old postdoc, she could not imagine waiting to have children until after tenure at age 40, nor could she imagine how she would juggle caring for a young family with the omnipresent demands of an assistant professorship." According to Williams and Ceci, "Jennifer made the choice to have a family and teach mathematics part time at a local community college."

Williams and Ceci argue, based on their extensive review of the psychological literature on the gender gap in science, that "This single factor of having or wanting children has great impact." How great is it? Williams and Ceci assert that "In some analyses, females are twice as likely as males to decide not to pursue tenure track careers as a result of this factor." They add that "Having children prior to working as a postdoctoral associate creates a 19 percent likelihood for men of opting out, versus 32 percent for women; having new children after beginning postdoctoral work creates a 20 percent versus 41 percent difference."

Williams and Ceci succinctly state the process by which the gender gap in science grows: "The percentage of women in the applicant pool at each of several key transition points" declines: "From award of Ph.D., to application for tenure-track positions, to being invited to interview, to being offered a tenure-track job, to being promoted to associate and full professor, women's numbers diminish." (p.143)

Although Williams' and Ceci's discussion of the impact on women of having a child concerned the situation in the United States a European government report makes abundantly clear the impact of having a child is not simply a dilemma for American women.

The European Commission (2003), in their report on women engineers, *Creating Cultures of Success for Women Engineers,* noted "For women engineers, maternity is seen as a problem and correlated with professional difficulties." The Synthesis Report asserted

> Companies use it to put pressure on mothers. Maternity seems to be only a woman's problem; women talk about their arrangements concerning children, as if it concerned only mothers, and the companies observed adopt the same discourse: maternity and its consequences only concern women as mothers." (p. 51)

In some societies within the European Community the employers are totally unsympathetic to women who become mothers. In Greece, "very often women would be the first to leave in case of economic lay off." In Slovakia, a respondent alleged that women, especially those who because of maternal duties could only work part time or needed special consideration were "the easiest victim" of any misfortune befalling the employer that necessitated layoffs. The Report concluded that "maternity is still a woman's affair, and represents an 'eternal dilemma' (in the words of a Slovakian respondent)" (p. 51).

While Williams and Ceci are correct that the single factor of having or wanting children" has great impact, the question arises as to just what is this single factor. Is it a psychological variable—as suggested by Williams and Ceci—or a complex set of both sociological and psychological variables, many of which are not well understood? I lean to the latter view.

To understand my position, it is best to begin with the evidence Williams and Ceci muster. They point out, for example, that a gap exists "between the proportion of applications for tenure track positions from women and the proportion of doctoral degrees awarded to women" and that this gap is "more pronounced in some fields than in others; for example, the difference is quite substantial in chemistry and biology, two disciplines with relatively high proportions (32 percent and 46 percent, respectively) of recent Ph.D.s awarded to women." They further note, "When women do apply for tenure-track positions, they are more likely to be interviewed and hired than their proportion in the applicant pool (as opposed to their proportion in the Ph.D. pool) would lead one to expect."

Williams and Ceci argue persuasively that it is "not biased hiring committees, journal reviewers and grant panelists [that] are the primary cause of the under representation of women in academic science." They rightly point out that "The economists Marianne Bertrand, Claudia Goldin and Lawrence Katz surveyed nearly 2500 University of Chicago M.B.A.s who graduated between 1990 and 2006." The researchers found that "More women M.B.A.s dropped out of the full-time workforce with each added year following their degree."

Then Williams and Ceci begin to make their naturist argument asserting that as the women's "*biological clocks ran down*, (emphasis supplied) significantly fewer" [continued to work full time]. Williams and Ceci added,

> Although this situation is also seen among women in medicine and among Ph.D.s, it is exacerbated in business, particularly for those jobs associated with workweeks of more than 60 hours (investment banking, consulting) or close to 60 hours (venture capital, sales, trading).

Are Williams and Ceci correct that the women's biological clocks were running down. The answer to that is maybe. The authors did not have medical information to justify that statement about the study sample but it is certainly true that in general women who are past thirty-one have a higher probability of bearing children with birth defects etc. Yet, what Williams and Ceci neglect to do is rule out any competing sociological hypotheses before proposing their psychological hypothesis.

An example of a competing sociological explanation is one that I proposed in a rejoinder: perhaps the women were entering the sandwich generation years and besides any problems with a ticking biological clock they also had parents with health problems who were now demanding more help from their offspring in their declining years. (See letter to the Editor, The American Scientist).

Rephrased in statistical terms, Williams and Ceci correctly see a correlation between women aging—and their fecundity declining—on one hand, and their labor force participation, especially full time participation, declining on the other. The question Williams and Ceci must answer is whether that correlation indicates causation or is merely spurious.

I maintain the correlation between women's declining fecundity as they pass their thirty-first birthdays and their leaving the work force (or reducing their hours) is spurious.

It is important to make two final observations about the thoughtful contribution that Williams and Ceci have made: first, no one has shown their views to be incorrect. I maintain, however, that before psychologists go and do studies to find the psychological drive to have children they rule out mundane sociological explanations behind the women's dropping out of academia. Thus far, all the most cogent explanation for women's dropping out more than men emphasizes the importance of such social factors as the overwhelming stress presented by parents needing assistance in their declining years and the adult offspring's entering a less fecund period of their lives.

Second, Williams and Ceci, to their credit, see a need for changes in academia to accommodate women's need to start a family. They offer some thoughtful suggestions for policy changes and practices that would allow more women to stay in academia.

I have completed my review of the "nature" theorists who, as the reader can see, come primarily from psychology and who argue for a biological basis for the social phenomenon of the gender gap.

Previous research (Etzkowitz et al., 2000; Fisher, 2010) and now the present study demonstrate by the time a white male student receives his degree, he has accumulated invisible but valuable assets that his female or minority student lacks—a circle of colleagues who themselves are working at the cutting edge and who help him throughout his career, an easy confidence that he belongs and does not have to prove it by meticulous planning and other time consuming work that lessens productivity etc. Women and male graduate students who are not white have fewer collaborators according to Bozeman and his associates (2003, 2004), and must follow more time consuming problem choice strategies in an effort to demonstrate they belong in science (Fisher, 2005).

The totality of subtle advantages white men enjoy have been beautifully captured in the memorable words of Jonathan R. Cole (1981, quoted in Fox, 1992) "although women have moved into science they are not *of* the community of science." He further noted that women more often than men remain "outside the heated discussions, inner cadres, and social networks in which scientific ideas are aired, exchanged, and evaluated." What Cole has pointedly said of women relative to men applies as well to Hispanic heritage and African ancestry students, and even to male Asian students in comparison to white male graduates—though perhaps to a lesser degree.

How do white male graduate students accumulate so many advantages? Etzkowitz, et al. (2000) provides the best description of this. They make it clear that the vital contacts, the friendships with potential collaborators and discussants are forged in the informal circle of students or coterie working under the direction of a senior professor. According to Etzkowitz et al. (2000), as they hang out together in the local watering hole, drinking a beer the coterie of graduate students around the senior professor talk about a variety of issues, sharing opinions about new research studies, new ideas for research, the virtues and vices of different researchers. They share intelligence both among themselves and with their dissertation sponsor who when they are ready to embark on their careers will refer them to his close colleagues and friends in academic institutions elsewhere. And the professor's close colleagues and friends will in turn introduce these junior scientists to other people who are doing important work.

It would be wrong to say none of the African ancestry and Hispanic heritage students benefit from this valuable, albeit, informal education. However, in general, the other students do not have the same easy camaraderie have with their professors—and perhaps with each other—as the white male students. Consider the observations of Etzkowitz et al. (2000) on this point. "Contrary to gender stereotypes, female graduate students are often left to be the 'rugged individualists,' having to fend for themselves" (p. 74; also Fisher, 2007, p. 28). And, they added, the "precarious status" of women graduate students can be seen in comments such as this one of a woman respondent: "I felt like I didn't have any backup support. I didn't know how to pick a topic [for the dissertation]. The guys talk about that at the bars. I don't go there." It can also be seen in the observation Etzkowitz et al. (2000) made that although informal study groups (e.g., for studying for the comprehensive examinations) are important for success in graduate school and beyond, "We identified some degree of exclusion [of women graduate students] from study groups in virtually all departments studied (p. 74)."

Leonardi (M.A. thesis, 2003) asserts that engineering students "used . . . communication skills to manipulate their peers." He illustrates his point by showing that study groups in engineering school are arenas for interpersonal competition for control. As an example of this manipulative nature of communications he offers the following response of one of his study subjects:

> When you form a group [for a class project] the first thing that happens is you have to decide who is telling who what to do. People have to know what they are doing and the only way you can do that is to have one person assigned that in the group. And so, the first that happens is that people are trying to become the person in charge (p. 67).

My research has shown that women are repelled by this bruising battle for dominance in the group. Yet in highly masculine environments—and many science and engineering departments are heavily masculine although women have begun to show up in engineering schools in significant numbers—this often occurs.

Exclusion from study groups (Etzkowitz et al.), persistence of student sexism among students in the sciences (see CATSG), extreme competitiveness among engineering students (Leonardi) and perhaps students in the hard sciences—especially Chinese students who are trained in highly competitive environments in their country (see sinostand.com)—are examples of the obstacles women face in graduate school that either drive them out of science and engineering or lead to their coming out with credentials that, on their face are equivalent but in fact are qualitatively inferior to those of their white male colleagues. This is the ineluctable implication of Bozeman's research, of Cole's observations (shared by Fox), and even of the findings of Ginther et al.'s work in regard to African ancestry researchers.

Beyond these advantages described in the literature that white male students enjoy, are still other advantages white male students seem to enjoy that I documented in my previous monograph: the white male students were less likely to complain that assistantships imposed heavy undergraduate teaching duties; more likely to say the school hired good professors, more likely to claim that the school brought in speakers to talk about opportunities in the field for people with terminal degrees, and more likely to give the students access to latest equipment for their research.

In sum, the qualitatively better graduate education the white male students receive compared to the other students—an education that confers subtle but significant advantages—shows up in the form of white male professors winning more grant support, quicker tenure and doing more and better publications in their careers.

Despite the less favorable treatment that minority students receive compared to white male graduate students during their graduate education, minority students continue to increase their enrollment in graduate programs. Data compiled by the National Center for Education Statistics and reported in Fast Facts (2011) show that "From 1998-1999 to 2008-2009, the number of degrees earned increased for students of all racial/ethnic groups for each level of degree."

Even more interesting is that "During the [1998-1999 to 2008-2009 period], the number of bachelor's degrees awarded to Black students increased by 53 percent, and the number awarded to Hispanic students increased by 85 percent."

In fact, according to the National Center for Education Statistics, "in 2008-2009, Black students earned 10 percent and Hispanics earned 8 percent of all bachelor's degrees conferred, up from 9 and 6 percent, respectively, in 1998-1999." Furthermore, the Center found that "higher percentages of master's degrees were conferred to Black and Hispanic students in 2008-2009 (11 and 6 percent, respectively) than in 1998-1999 (7 and 4 percent, respectively)."

The National Center for Education Statistics also looked at doctoral degrees conferred, and they found more African ancestry and Hispanic heritage recipients in 2008-2009 (4,434 Blacks and 2,540 Hispanic doctoral degree recipients versus 2,136 Blacks and 1,302 Hispanics in the 1998-1999 period. That is a more than doubling of the doctorates awarded in the later period compared to the earlier period. It is simply absurd to say that African ancestry and Hispanic heritage people have no aptitude for science in the face of these statistics.

These data should put to rest any notion that African ancestry and Hispanic heritage students do not have the ability to do science. After all, while African ancestry and Hispanic heritage students are a minority of the graduate student population receiving degrees in the STEM disciplines, more Americans of African ancestry and Hispanic heritage are attending and graduating from college and graduate schools than did ten years earlier. Would this occur if African ancestry and Hispanic heritage students had no talent or interest in science because of their genetic endowment?

I need to address one further issue before turning to a discussion of the theoretical perspective driving the current study: how much did the African American population increase in the same period as the increase in degrees awarded. Did the African American population in the United States double in that ten-year interval along with the increase in degrees awarded? That seems highly unlikely but why should we appeal to plausibility when we can put the matter to rest with the data? NACME says in its 2011 Data Book that

> African Americans account for 12 percent of the U.S. population and will remain at that proportion through 2050. . . . Latinos now account for 16 percent of the U.S. population but will grow to represent nearly one-in-three Americans by mid-century. . . . In total, members of the three underrepresented minority categories (African Americans, Latinos, and American Indians) constitute 29 percent of the U.S. population now and will account for 43 percent by 2050).

If the literature is so compelling in its documentation of the superiority of the informal education white male students receive in graduate school compared to others, why the need for the present study? I can give two reasons here for the need for this study. First, this study seeks to test a hypothesis that Cole and Singer (1991) advanced regarding why women and minorities such as African Americans, Hispanic heritage scientists, and American Indians are promoted more slowly, receive tenure less frequently, and are underrepresented in the ranks of senior professors in academic departments in the STEM disciplines. The hypothesis argues that prior socialization is behind these findings of a gender gap. Second, this book intends to throw light on the situation of African ancestry and Hispanic heritage graduate students in particular which may account for their relative absence from the ranks of faculty in the STEM disciplines.

It is now time to turn to a brief presentation of the "limited differences" theory of Cole and Singer (1992), the sociological theory that I see as generally explaining the gender gap in science.

The first point I want to make is that Cole and Singer's theory rests on a view of human psychology just as the "naturist" theories do. However, whereas the naturists see a biological imperative Cole and Singer offer a "social learning" perspective—especially the ideas of Albert Bandura (1977). Central in my opinion, to the theorizing of Cole and Singer is the idea that women are socialized to depreciate their talents as scientists. Their "theory of limited differences" points to women's reactions to slights as well as the slights themselves, that lead women to want to leave science. I suggested that since implicit in their theory is that women's experiences in graduate school might be important to women dropping out of science graduate school was a strategic site for assessing their theory. In fact, I found that graduate school experiences—the sexism of some fellow students, the practice of foisting heavy responsibility for undergraduate education on graduate students who held assistantships, lack of encouragement to start or continue research, all were contributing to women feeling more isolated and discouraged in graduate school than white men. In short, I believed that I found important support for one central piece of the Cole and Singer model.

Cole and Singer acknowledge that graduate school affects women. However, they insist that previous socialization also is extremely important. First, of course, it leads to leaving the pipeline for science and engineering prior to graduate school. It also may make the women more vulnerable to pressures to leave once they enter graduate school.

I must emphasize that a proper test of the Cole and Singer theory would be a very expensive and long-term cohort study lasting perhaps forty or more years. This is unlikely to be done in a time of fiscal austerity. However, a crude test of the ideas in their theory is possible. One way to do this is to say that if you have people with very different backgrounds, e.g., African ancestry students, white students, Asian students, etc. and give them the same treatment (e.g. graduate schooling in the same discipline) you in effect are doing a natural experiment that evaluates the impact of the background variable of ethnicity (which is a kind of socialization variable) on the students. In my current study this was my strategy. However, I did not implement my strategy as elegantly as I would have liked First of all, the "graduate school" treatment was not exactly identical for each group. In the case of the African American and Hispanic students it was the 60 prominent research universities I sampled to get these students. In the case of white men, women (other than African ancestry and Hispanic women), and Asian students it was the schools in three regions of the United States. (This variation in the treatment was the intellectual basis for my critic's accusation about my study design being "an act of desperation"). However, the results, I believe, are still instructive in showing how ethnic background interacts with education to determine attitudes and perceptions. In short, the data, while imperfect, are useful for policy purposes in urging certain practices and policies in graduate schools be modified.

On the other hand, in my earlier monograph (CATSG) I did not delve into school atmosphere in anywhere near the degree that I think it deserved. In the present monograph I take up the subject once again and devote an entire chapter (Chapter Four) just to the relationship of manifestations of sexism (e.g., attitudes that men make better scientists, professors, and team members) to a school atmosphere where professors are not respectful of gender or ethnic differences. The findings show that school atmosphere can make a difference in the sexism that even women manifest, e.g., women believing men make better scientists. I maintain that this sexism makes these women more vulnerable to pressures from spouses and others to forego a career in science. In other words, prior socialization—attitudes women may have before they enter graduate school that they are somehow not as good as men or not as good as white men—may contribute to dropping out just as Cole and Singer would have expected.

The other major contributor to my thinking about why women are not in the top ranks in science is Etzkowitz's views about professions and how they perpetuate themselves. The essence of their thinking was presented in the Introduction and will not be repeated here.

It is useful at this point to summarize the principal hypotheses driving this study based on the literature review.

1. The "reproduction of the profession occurs in a way that selects for people with similar social, cultural, and economic characteristics to those already in the profession' (see Making Science Fair, p. 28; originally quoted from Etzkowitz et al., p. 52)."

2. As Cole and Singer maintain, greater research productivity leads to more resources for research, quicker promotion, quicker receipt of tenure, and so on. Women do not reach tenure as quickly as men, nor receive promotions as quickly as men because their prior socialization does not prepare them as well as men for the world of the university. Women come to the starting line with various disadvantages from their prior socialization, including (as I demonstrated in CATSG) (1) fewer contacts and collaborators for research, or (2) dissertation sponsors who were not as prestigious as the men's sponsors (suggested by Cole and Singer) and so on.

Whereas Etzkowitz et al. place great emphasis on the graduate school as the primary source of the gender gap that occurs in scientists' subsequent career, Cole and Singer diffuse the responsibility for this phenomenon. While Cole and Singer argue implicitly that graduate school is less advantageous for women than men, they also emphasize that women's disadvantages begin much earlier and are cumulative. For example, Cole and Singer see the high school experience of women and their college experience perhaps contributing to women's aversion to science and mathematics as career choices. For women, and perhaps for African ancestry and Hispanic heritage students of either gender, it is the accumulation of disadvantages prior to their reaching the starting line of their careers in science, engineering and mathematics that Cole and Singer emphasize is responsible for their lesser success in science. This emphasis on the lifetime accumulation of disadvantages relative to men constitutes the distinctive feature of their theory.

IMPLICATIONS OF HYPOTHESES DRIVING THIS STUDY

I want to discuss in turn the implications of each of these broad hypotheses for the present research study. I will address each in the order presented.

A profession that seeks to reproduce itself has numerous means to ensure its survival, some of them perhaps legally proscribed but others permissible. For instance, in the United States a profession cannot legally tell persons of one ethnic or racial category or one gender not to bother applying for entry to the profession's training programs. The Constitution of the United States and other state and federal laws prohibit that.

On the other hand, I maintain that this does not prevent the profession from taking other steps to limit entry to those whom it wishes to join the profession. For instance, it is not illegal for white male professors to informally encourage white male students to start or continue research in the context of informal "get togethers" at the local pub. Women especially those whose cultural and religious traditions look askance at women patronizing pubs will not be present at these gatherings and will not get that extra encouragement.

Nor is it illegal for a university to use teaching assistantships to finance student enrollment although a clear pattern of giving them to women or to minorities while awarding fellowships to men, especially white men, may be grounds for a class action lawsuit. Still, the affected students first have to realize that they are being discriminated against, and then decide whether they wish to fight for years through the courts for justice. It is not hard to sympathize with students who having weighed their options simply elect to get out and go to professional schools in law or medicine or business instead. My guess is that in the vast majority of cases the women and minority students will opt for that professional school solution rather than choose to stand and fight to reform a clearly discriminatory graduate school program.

Unfortunately for the nation as a whole, the departure of these students from the STEM disciplines may ultimately be a loss for the country's scientific prowess. Their departure cannot have come at a worse time in my opinion. Thanks to competition from abroad (from economic powerhouses such as China, India, Brazil to name the most notable) and difficult challenges in environmental quality, slow economic growth with high unemployment, declining health of an aging population, transportation infrastructure needs, etc. we need more highly trained scientists and engineers to help us innovate our way out of our various social problems.

If Etzkowitz et al. emphasize the machinations of a profession seeking to reproduce itself by recruiting people much like the ones in the profession currently, Cole and Singer emphasize the reasons women begin exiting science even before they go to college. While they may have the intelligence to do science, may in fact be quite proficient in mathematics, Cole and Singer argue the women become discouraged by assorted slights, insults, and setbacks and they abandon the field. Women begin leaving the pipeline for careers in science perhaps as early as high school. It definitely occurs in college. For those who survive college with their interest in science intact, the decision to quit may occur while in graduate school. And, as Williams and Ceci insist, it also definitely occurs not long after the women achieve their doctoral degrees.

Up to now I have been laying out my position at a rather general and abstract level, not at the level where research can be accomplished. I need to translate general concepts such as "reproducing itself by selecting" into specific testable hypotheses.

I will begin, however, with a general observation. If a profession wants to select leaders of the profession similar to those it already has, it stands to reason that it wants to recruit people similar to those already in the profession in regard to values that are important. It is most comfortable with people who have more in common with its current members than with those who have less in common because it is easier to believe that people who have much in common with you also share the values that you regard as important. It is this desire to assure that the future leaders of the profession continue to uphold the core values of the profession that drives much of what the profession does.

The fear that women, or that African ancestry and Hispanic heritage people, do not share those core values, I maintain, lies behind the various strategies a discipline uses to resist opening its top ranks to women and minorities. And the strategy that seems to have arisen in the university setting is to create a chilly climate within its training programs for women and minority people seeking admission to the ranks of the disciplinarians. I intended to document this chilly climate with two kinds of evidence. One is evidence of relatively less encouragement of women and minorities in some disciplines than in others. According to this line of thinking certain relatively open professions such as biology, social/behavior sciences and other "soft" sciences are more accepting of women in all ranks than are their "hard science" sister disciplines. A second kind of evidence is the evidence that some segments of the graduate student population claim men make better scientists, professors, than women. A third kind of evidence I considered was more general "atmosphere" variables—whether the students feel a school is "snobbish" or "intellectual" or "competitive".

Since I place great emphasis on the encouragement students feel in their educational experience in making my argument as to the underlying causes of women and minority students dropping out of graduate school and/or not attaining high status in the discipline, I need to amplify what I mean by "feeling encouraged." Encouragement can take two different forms: for instance, direct encouragement and indirect encouragement. Professors can informally encourage students to start or continue research in the discipline, communicating their confidence the student is a good prospect for a research oriented discipline. Schools can award the students financial assistance communicating their belief the students have aptitude for the discipline. These are direct forms of encouragement. Indirect encouragement can take the form of fostering a general atmosphere in the school that seems congenial to the students. Students generally prefer a "friendly" atmosphere as opposed to a "snobbish" atmosphere. Women in particular dislike a snobbish atmosphere that smacks of efforts to stifle their opinions and ignore their feelings.[1] A competitive atmosphere is desirable from the students' standpoint but not one that is overly competitive—which is often what occurs in a school with a snobbish atmosphere.

In summary, my specific hypotheses are as follows: White men students will feel more encouraged to pursue a doctoral degree than African ancestry students and Hispanic heritage students, as demonstrated by:

- more often reporting that the school hires good professors than other component populations, especially minority students

- less often reporting that assistantships have burdensome undergraduate educational responsibilities
- more often reporting access to the latest equipment to do their research
- more often reporting that the financial aid is satisfactory
- more often reporting that their professors are respectful of gender and/or ethnic differences
- more often reporting that they were given opportunities to work on problems as graduate students

The second general hypothesis I consider to be rigorously derived from Cole and Singer's work. Cole and Singer did not point explicitly to differences across cultures in how women might suffer disadvantages relative to men that affected their careers in science. They did not ever say, for example, that Iranian or Yemenite women might be less prepared to do science at the frontiers, not only compared to Iranian men and Yemenite men but also compared to native born white women in the United States. However, they would acknowledge that if the evidence were adduced that Yemenite women and Iranian women faced strong pressures to be wives and mothers from their spouses, their parents and others compared to other women certainly then the Iranian and Yemenite women would be less likely than women of other cultures to seek careers in science that interfered with the women's wifely duties. In other words, their model would be consistent with such a hypothesis although they did not explicitly offer this hypothesis. In contrast, the hypothesis of the resource deficiency school (generally identified with writers such as Mary Frank Fox (1992; 1995) or Yu Xie and Kimberlee Schauman, 1998; 2003) would not be able to address this issue at all because it does not consider prior socialization in explaining women's careers in science.

What would be the dynamic in the model I propose to predict that Iranian and Yemenite women are more likely to avoid careers in science than white women born in the United States? I maintain the dynamic would be the sexism that Iranian and Yemenite women have internalized.

The Iranian and Yemenite women either would have been socialized to believe science work was men's work or, if they really enjoyed science work, they would have come to believe they are not good enough to do science at the cutting edge; and furthermore, that they were meant to be wives and mothers and perhaps work part time to help the family financially.

Graduate students from cultures where women are expected to be deferential to men will show more sexism than those from cultures that do not expect women to be deferential. For example, students from Asian cultures, regardless of gender, will be more sexist than native-born white students.

Sexism of students is linked in my opinion to the limited numbers of women in the graduate school either as professors or as students. If I am correct, the sexism will not disappear anytime soon, especially in the hard sciences. Robin Wilson, writing in The Chronicle of Higher Education on

February 16, 2012 remarked, "It could take nearly 100 years before half of all professors in science and engineering are female." He quotes Deborah Kaminski, who along with Cheryl Geisler, wrote an article published in Science in February 2012 with projections of future numbers of women science and engineering professors. According to Wilson, in their article Kaminski and Geisler estimated "it may be 2050 before 50 percent of new hires in science and engineering are women" and "at least another forty years before the actual population of science, engineering, and mathematics professors is 50 percent women."

These data are consistent with my own expectations that as much as thirty or more years may be required to bring women physics faculty up from perhaps four percent to around ten percent of the total in the United States. I based these results on experience in other countries, notably Denmark but also other European nations that began pushing for more hiring of women science and engineering faculty back in the 1970s and usually took decades to double their women faculty in various disciplines. (See Langberg, Kamma, 2005; also Fisher, 2007, p. 34 and p. 95.)

The barriers to increasing women faculty in science are various. One is the simple fact that professors do not have to retire and make way for younger scholars. Professors can stay at their posts well into their eighties. I personally —emeritus professor—and is past her hundredth birthday.

It is one thing to dismiss inherent intellectual deficiencies in African ancestry and Hispanic heritage students as the basis for their near absence from the ranks of top scientists. It is quite another to identify what specific sociological variables may be significant in explaining the absence of sizable numbers of people of these ethnic backgrounds from the ranks of top scientists. We do not get very far in saying the reason for their limited numbers is that they have the misfortune to live in a sick society that fails to adequately utilize the talents of all its citizens. If we are to correct the problem we need much more information on what the illness is that afflicts the body politic.

I offer such a hypothesis taken from my previous work on the under representation of women in the top ranks of science: the "reproduction of the profession occurs in a way that selects for people with similar social, cultural, and economic characteristics to those already in the profession' (see Making Science Fair, p. 28; originally quoted from Etzkowitz et al., p. 52)." However, some professions have been noticeably more welcoming of women and minorities than others, which would not necessarily be suggested by the basic hypothesis. Professions are open systems interacting with their environments. Nursing, although it was men's work in its beginnings became an occupation that women soon began to enter in large numbers and more recently has again attracted men to its ranks. However, it remains a field relatively welcoming of women. In contrast, as indicated in the literature certain academic disciplines have been less welcoming to women and under-represented minorities than others. For example, certain specialties in engineering and physics, have resisted penetration of their top ranks by women far more than biology and psychology.

DATA COLLECTION

African ancestry and Hispanic heritage students clearly respect the value of a post secondary school education. However, it is true they are not showing up in graduate schools in the science, engineering, and mathematics programs in numbers reflective of their proportion in the general population. The data I quoted from the National Center for Education Statistics offers hope that eventually this situation will change.

The number of people of African ancestry and Hispanic heritage in the future who will seek doctorates surely will rise even if as a proportion of the African ancestry and Hispanic heritage population no change occurs. The reason is simple. Both in numbers and as a proportion of the total, Hispanic and African ancestry people are a substantial and growing component population in our nation as I pointed out above (supra).

NACME added, "The Latino population is the fastest growing segment of the total U.S. population." NACME argues, and I agree, that

> The solution to America's competitiveness problem is to activate the hidden workforce of young men and women who have traditionally been underrepresented in STEM careers—African Americans, American Indians, and Latinos.

A key aspect of activating the hidden workforce is encouraging more of them to pursue advanced education in the sciences and engineering. If our country is to meet the shortage of highly trained people in the STEM disciplines in the future we must inevitably tap many more African ancestry and Hispanic heritage individuals. It is a simple matter of demographics, especially as reduced fertility among white people in the United States and the changing interests of white males—from whose ranks traditionally graduate schools recruited most of their students—lead to lower enrollments of these students in graduate science and engineering programs in the future.

The demographic changes in the American population—declining fertility of white people, increasing numbers of Hispanic heritage and African ancestry people, make it essential that we attract more Hispanic heritage and African ancestry graduate students to the sciences and engineering programs. I maintain that to do so we need information on what is attractive and what is not about these programs from their standpoint. However, regardless of the reason African ancestry and Hispanic heritage students are not showing up in graduate schools in the S and E disciplines in large numbers and it poses a practical problem for research on these students.

As I noted earlier, in my most recent monograph Crippled at the Starting Gate I was able to identify just sixty such students out of well over a thousand respondents in my sample. That is five percent! For the present study, I was able to find 240 more such respondents to add to the 60 Hispanic heritage and African ancestry graduate students in the STEM disciplines I already had. I accomplished this by scouring sixty of the major research universities nationally—but it was hard work to find suitable respondents.

Research Procedures

Definition of terms

Up to now, I have been talking about "African ancestry" and Hispanic heritage people as if everyone knows what I mean. That was acceptable in my discussion of why I wanted to do a study of African ancestry and Hispanic heritage graduate students. It is not sufficient for planning a survey. Operational definitions of these terms are necessary and that actually is a knotty issue. However, the time has come to say exactly what I mean since I intend to research their attitudes for a study of policies and practices of graduate schools that need remediation.

Who Should Be Studied?

Many concepts in social science have exact meanings such as gender or chronological or income category (in some standard unit of currency such as U.S. dollars). For others however it is more difficult to make operational definitions that all can accept. Consider the question, do Jewish people support Israel's position on how to make peace in the Middle East? Surely, all can agree that this is a question of some policy significance. But to research this issue is not so simple. For example how do we define a "Jew?" I consider it a religion. On the other hand, according to the New York Times issue of 14 January 2006, (cited in Wikipedia), scientists have "found Ashkenazi Jews (i.e., Jews whose ancestors lived in Central and Eastern Europe in countries such as Germany, Lithuania, Latvia, Poland, the Czech Republic, Hungary, Russian, Ukraine) to be a clear, homogeneous, genetic subgroup."

In other words, Judaism is not simply a religion if these scientists are correct but a people with genetic differences in common with one another that differentiate them from non-Jews. Throughout history, their enemies have treated Jews as a people. The Nazis for example saw the Jews as a people who needed to be exterminated. People who were of Jewish ancestry even though they were not Jewish were unfit to be members of the Nazi party.

The foregoing (and somewhat tangential) discussion of "Who is a Jew") contains an important general point. Just as deciding who is a Jew has important consequences, deciding who is person of Hispanic heritage and who is a person of African ancestry is a decision with consequences. In other words, any definition is likely to be open to challenge not just from fellow researchers but also from stakeholders.

One way out of the dilemma is to note that even if there is no universally correct definition, it is possible to arrive at one that is appropriate for the research study one wishes to do. In other words, perhaps I could state a reasonable definition if I could satisfactorily answer the question: what is the purpose of my study? I decided the purpose of my study was to provide information to professors and deans—in short, to policy makers in graduate schools—to help them make graduate schools more welcoming environments to people who were not white males.

Knowing what my study's intent was—provide policy makers with ideas on how to make the schools more welcoming to students who were not native white males—proved decisive in helping me answer the question for this study: whom do I count? Specifically, who is the African ancestry student that I must include in my study? Who is the Hispanic heritage student I should be including in this study? I already had input from women and Asian students on what aspects of the graduate school experience were especially bothersome to them and perhaps could be remediated. Now I believed that given the changing demographic situation of the United States, with declining white fertility on the one hand and higher fertility of the African and Hispanic component populations on the other, graduate schools needed to recruit and select students from these latter ethnic groups. And the policy makers needed to make the schools welcoming for the Hispanic heritage and African ancestry students as well.

Did this mean I should have a group of respondents who resembled the likely demographic mix of the Hispanic community in the United States? Did it mean I should choose African student respondents who were representative of the African American community? Dr. Michael Maranda, a researcher on the staff of the City University of New York, who had done considerable research in ethnicity and to whom I turned to for advice, argued cogently that the relevant group of Hispanic graduate students would be only people of Puerto Rican and other backgrounds that were substantially represented in the population of the United States. He would have answered in the negative the question: is a person of Italian ancestry born in Argentina a Hispanic heritage person? He similarly argued that foreign trained African students should not be lumped together with the African American students in the sample insisting that the literature (e.g., Mary C. Waters, 1999) suggested foreign born African ancestry students were likely to be better treated than African Americans.

I chose to largely reject Maranda's advice in pursuing this study. If the aim were to make the schools more welcoming to people other than native white males, I could accomplish this by taking an expansive view of African ancestry and Hispanic heritage. After all, I reasoned, America is still a nation of immigrants. It is true most of the Hispanic immigrants are from Caribbean island nations such as Puerto Rico, Cuba, Dominican Republic or North American and Central American nations such as Mexico, El Salvador, and Guatemala. But should I not count Chileans or Venezuelans? I decided that American graduate schools should be welcoming people from anywhere if they can contribute to our society. And how better to make them welcome than to find out what they like and do not like about graduate school in the United States in the sciences and engineering?

I do not wish to suggest that I took the position I did because I am a great humanitarian. The simple fact is I had great difficulty finding African ancestry graduate students and Hispanic heritage students even when I took a broad view of whom to include. I might have had to abandon the study altogether—Maranda's preference—if I insisted on choosing only students who reflected the current demographic characteristics of the American population.

Simply because I decided on an inclusive view of suitable respondents did not mean I had no challenges to address. I might argue logically that Argentine birth could be considered evidence of Hispanic heritage. On the other hand, if the person had an Italian family name such as Marconi but had been born in Argentina, where the residents include a high percentage of people of German and Italian ancestry, should I include or exclude that person? I opted for including them if they claimed to be Hispanic heritage. However, I excluded Brazilians, Cape Verdeans and others of Portuguese cultural background.

I preliminarily accepted as being in the African ancestry group some who said they considered themselves "Sudanese" or "Guyanan" if I felt they were likely to be African heritage. Many Guyanans are of Indian heritage and would not qualify as African and, since they speak English, would not qualify as Hispanic heritage either.

While I have implied it, I have not come right out and said that I depended on self-identification in selecting "African ancestry" and Hispanic heritage students. Not every researcher will be pleased. However, I am not the only one to follow this convention. Although the United States census employs a more elaborate procedure to make the decision as to what ethnic/racial group to place the person in it is still a choice of the respondents because the respondents first decide whether they consider themselves culturally Hispanic. Then, regardless of how they answered that query they decide what race they consider themselves, including mixed race.

I elected to do a much simpler classification for a small survey over the Internet. But fundamentally, both the census and this survey had more in common than not because in both cases the choice of what ethnic/racial group the respondent belonged in was to a considerable extent up to the respondent.

Since I framed the question regarding ethnic/racial group affiliation as an open-ended one, students who referred to themselves as "Afro-American" or "black", I preliminarily accepted as being in the African ancestry group. However, as I noted above some said they considered themselves "Sudanese" or "Guyanan" or "Cape Verdean" as their response to the open-ended question.

How do you count a person who calls himself "Sudanese" or Guyanan? How do you classify a person who claims to Afro-Caribbean and states he or she was born in Cuba?

For all cases I was not absolutely certain what to do based on the open-ended question answer, I decided that to use further screening criteria to decide how to categorize them. One important criterion in making decisions about who to include among African ancestry people was the respondent's and his/her parent's place of birth.

The classification process I followed produced a group of respondents that included substantial numbers of foreign born "African ancestry" students. Despite Maranda' s criticism I decided to include these students in my study population. Not only that, I decided to combine the foreign born and the native born African ancestry students in one group and the foreign born and native born Hispanic heritage students in another group. The reader is entitled to understand my basis for these choices.

First, I felt strongly that the foreign-born African ancestry graduate students who took the trouble to answer my questionnaire deserved to be counted. Second, these foreign-born African ancestry students had perceptions that need to be reported and need to be taken into account in designing reforms of the graduate school experience to make the graduate schools more welcoming to students who were not white males.

These sentiments tipped me in favor of a more inclusive view of "African ancestry (e.g., more respondents equals richer information). However, I still needed to address the possibility that my sample was so unrepresentative of the American demographic situation that I would be providing erroneous policy advice for my audience of professors and deans.

I needed some persuasive reason for continuing the study analysis and reporting the results and that would have to be a finding that counting the foreign born Africans in the group of as "African ancestry" would not distort my policy advice.

I chose to do an empirical test as the basis for my decision: I asked an empirical question of the data: Could I differentiate foreign born African ancestry students from native born African ancestry students based on their answers to the policy questions that I investigated in my previous monograph? I reasoned that since I had small numbers even for this study, I would have to forego doing a quantitative analysis if the answer were affirmative. On the other hand, if it turned out that on the policy questions my foreign born and native born held similar views, then I believed for this analysis I could combine them and compare them with other large groupings such as white males and Asian males. I might also be able to treat African ancestry women and Hispanic women as women students rather than as a separate or two separate tiny categories.

In sum, my decision about what to do hinged on the results of a statistical comparison of native-born and foreign-born students. Tables 2.1 through 2.9 report the results of my comparisons of foreign-born African and Hispanic students with native-born graduate students. In no case did the differences meet the .05 criterion conventionally used to test for differences between samples. This decided the issue for me—I would count all the foreign born and native-born "African ancestry" students as a homogeneous group for my purposes.

One of the interesting results of my inquiry is that sometimes the differences, even if not statistically significant, were contrary to the direction expected if Maranda were correct. He would not have predicted that just 6.5 percent of foreign born versus 44.4 percent of native born African ancestry students were receiving financial aid because he read Mary Waters' work as predicting better treatment of foreign born. (Waters actually did not see foreign born as getting better treatment so much as she saw them trying to differentiate themselves from native born African ancestry people in an effort to obtain better treatment. Obviously, in the case of graduate school financial aid, claiming to be foreign born and not a citizen of this country was not an effective strategy for a student to follow.

Of course, remembering that Americans who have prior military service—as many African Americans do but few if any foreign-born ones would—helps explain why I found these results. It is commonly known that veteran's benefits help many people pay for their college and graduate school educational expenses.

The reader can see that I tried to decide the issue by weighing the benefits of a more inclusive view of "African ancestry e.g., more respondents and richer information against the negatives—the possibility that the views of my respondents were so unrepresentative of the American demographic situation that my study would provide poor guidance for my audience of professors and deans. While I acknowledged my sample was not representative of African Americans graduating with science and engineering degrees from American undergraduate educational institutions, I rejected the idea that my sample data would be useless for policy makers.

In other words, although Maranda was completely correct that my sample did not reflect the American demographic picture, and although he might likewise correct that in many respects foreign born African ancestry people receive better treatment in the United States than native born African-Americans, I believed that for purposes of this study my sample was large enough, and its findings persuasive enough about what African ancestry students see as areas in need of remediation to make the schools welcoming, that this would be a useful contribution to the literature.

While I obviously regard my decision to include these foreign born students as the right one, certainly the preferable one for my research purposes, I also believe that studies need to be done that are consistent with the advice Maranda offered since my study choices may not be correct for all purposes.

The issues that Maranda and I debated are issues that affect any survey. A survey must have both internal and external validity. External validity is the knottier issue in many cases than internal validity. It is knottier because the researcher must decide sometimes if the study should proceed with its external validity open to challenge.

This is an appropriate point to offer the following caveat about my study. While it provides the first quantitative data on the experiences of African ancestry and Hispanic heritage graduate students in the STEM fields, it certainly does not claim to be the definitive or final word. More important perhaps to the university official needing to determine policies and practices at his or her university the method I used to collect my data does limit the value of the research for policy purposes. African ancestry and Hispanic heritage students were chosen from among 60 research-oriented graduate schools. Obviously, this is not a representative group of all universities or even all universities that consider themselves "research oriented." Second, the experiences reported by the students were compared with those of white male students and Asian male students (as well as white and Asian women) surveyed some years earlier who had been chosen from a population of graduate students at three sites in the United States. I do not regard this unconventional approach as a fatal mistake procedurally or I would not have bothered to do this research. However, I certainly need to remind university officials reading this monograph that the most prudent course for those wishing to change school practices and policies to make them more welcoming to African ancestry and Hispanic students would be to study their schools first.

In this study I confronted the question: to whom can I generalize the findings of this study? I cannot claim that the results of this survey apply if the graduate schools suddenly admit a group of Puerto Rican and Mexican ancestry students with weak science training at the undergraduate level. If NACME is correct then it is plausible that the modal Hispanic heritage student graduating from college in 2012 might be rather poorly prepared based on their coursework and grades to handle the demanding courses or do the quality of research expected by the graduate schools. Certainly, they are probably not nearly as well prepared for the rigors of graduate school as the Hispanic heritage students in my haphazard sample.

On the other hand, I doubt the graduate schools will ever willingly admit large numbers of students with weak academic preparation. The schools will always search for those with a good undergraduate education because solid undergraduate preparation is essential to have a reasonable probability of being able to handle the demands of graduate school coursework and research. I am comfortable that my data therefore have some value for graduate school professors and deans in describing what they can do to make the school experience more fulfilling educationally for students other than white males who might wish to choose graduate school in the sciences and engineering in the future.

ENDNOTES TO CHAPTER ONE

1. Though the tables are not shown here, I infer women's distaste for a snobbish atmosphere from the following data. A substantial majority of male graduate students (61.7 percent) agreed or strongly agreed they felt encouraged to continue or start research for their doctorates in schools having a snobbish atmosphere whereas in schools not having a snobbish atmosphere only a slight majority of men claimed to feel encouraged (50.8 percent). Among women on the other hand, 46.0 percent in schools with a snobbish atmosphere felt encouraged to continue or start research whereas in schools without a snobbish atmosphere 56.2 percent felt encouraged to continue or start research.

Chapter Two
Cosseted White Males Revisited

Male and White-You're All Right
Hispanic or Black-Step Back

In this chapter I present my analysis of the views of the study population, the three hundred students of African ancestry and Hispanic ancestry who are the main concern of this survey. To some degree, this chapter represents a correction of the chapter reported in the previous volume. It is a correction in the sense that previously I had few cases of African ancestry and Hispanic heritage male graduate students on which to base any conclusions. Now I have a more adequate number. However, my enthusiasm is tempered by the realization that the Hispanic heritage and African ancestry students come from a different population frame than the others with whom I am comparing them. I do not think this is a fatal defect but it is necessary for readers to keep this in mind as they review the findings.

The present chapter illustrates the main aspects of the plight of African ancestry and Hispanic heritage males by pointing out white men's comparatively superior experience in graduate school. Compared to African ancestry and Hispanic heritage men, white men receive more encouragement to be successful in their graduate school, to finish their degrees, and embark on their careers.

It should come as no surprise that white men receive more encouragement to succeed in graduate school than other segments of the student population. Men traditionally have dominated the student ranks in graduate schools within the STEM (science, technology, engineering, and mathematics) disciplines and they continue to do so today (see Fisher, 2007; also NAS and other sources cited in Fisher). Men's overwhelming dominance in the ranks of scientists with advanced degrees is simply a consequence of the fact that men have historically dominated universities. These centers of higher learning first appeared in Europe nearly a thousand years ago at a time when women were barred from attending schools. In the nineteenth century the first American universities began to appear and they followed the German custom of barring women from matriculating at the school. Most also discouraged, if not explicitly barred, minorities including African ancestry, Hispanic heritage, American Indian, even Jewish students, from attending.

Women began to study in the graduate departments of universities in small numbers in the early decades of the twentieth century. Even today in some fields such as physics few women have achieved advanced degrees and still fewer have risen to the top of their discipline, although they have no legal barriers to overcome.

Because men have dominated in the university environment, the environment has become congenial for them. Universities only belatedly have begun to adapt to the presence of women in significant numbers. While universities are now adding staff to deal with student emotional health issues including drug abuse, suicidal tendencies, and other issues, these are recent developments. In the late 1990s when Etzkowitz et al. were doing their fieldwork, graduate schools that addressed the emotional needs of women graduate students were unusual if not rare. Until recently many universities have gone no further in their planning for women on campus than to create separate men's and women's restrooms. Yet it is clear the informal support network men can depend upon is not really available to women. As Etzkowitz et al. observed, quoting one women graduate student, "I felt like I didn't have any back-up support. I didn't know how to pick a topic. The guys talk about that at the bars. I don't go there" (p. 74).

Before I present my analyses, it is important to clarify what I mean by "encouragement." The concept of "encouragement"—or relative lack of it—of students is fundamental to my thesis that white men are more successful in graduate school because they receive more of it compared to less-favored graduate students.

Encouragement can take different forms, for instance, active encouragement. Professors, can informally encourage students to start or continue research in the discipline, communicating their confidence the student is a good prospect for a research-oriented discipline. Schools can award the students financial assistance communicating their belief the students have aptitude for the discipline.

Just as schools and professors can directly encourage the students, they can also indirectly encourage or discourage the students. They can create an atmosphere in which students of one gender, or one ethnic/racial group, feel appreciated, wanted, valued—or not. Departments and/or professors in those departments (such as physics) can do this for example by subtly (or not) communicating that for example women really are not likely to be successful in the discipline. Students will pick up the cues and reflect the attitudes of the professors. If the professors are skeptical women "belong" in physics, the students (certainly the men but maybe also the women!) will show they do not want women students as team members on multi-person projects, do not think women make good scientists or good professors. That is, the students' views however crudely mirror the attitudes of the faculty of their department. The faculty can intervene by taking steps that clearly undermine the view women do not belong: for instance, they can require teams to be mixed gender on all projects, seek women professors at the senior level, and encourage students to take a course in history or sociology of science where the role of women in science is addressed in detail.

Although indirect encouragement is important in my opinion, I leave it for another chapter (see Chapter Four). In this chapter, I confine my discussion to direct encouragement. After comparing the overall situation of African ancestry and Hispanic heritage male students to white male students, I turn to differences between disciplines.

Previous research has demonstrated women have had less success in reaching the top echelon of some fields such as physics and engineering than others such as biology, sociology, and psychology (see e.g., Fisher, 2007; Xie and Schauman, 2003, 1998; and Valian, 1998).

Minorities are even less visible at the top echelons than women. Is this because white men are encouraged more in graduate school than minority students? To help me find out, I decided to look at graduate school experiences of African ancestry students and Hispanic heritage students in graduate programs in the "hard sciences" versus the "soft sciences." This somewhat crude characterization of disciplines was all that I could feasibly accomplish given the size of my sample.

This chapter reports results for five groups of graduate students I believe are important for planning purposes: white men, Asian men, African ancestry and Hispanic heritage men, and all women. Except for the presence of Hispanic and African heritage males, the tables are similar to those in the previous monograph Crippled at the Starting Gate (which I will usually refer to as CATSG). I will depart from convention that calls for analyzing the data fully because most of the analysis of the new tables has already been done in the previous monograph. Instead I will emphasize comparisons between the African ancestry and Hispanic heritage graduate students on one hand and the white students on the other.

Sometimes, however, I will discuss (1) the regional versus overall white scores with an emphasis on their similarity (except where they differ by more than a few percent) and (2) the white versus African ancestry and white versus Hispanic heritage scores and (3) except where they differ meaningfully the lack of much variation in the scores of native born blacks versus foreign born blacks. The latter I see as one of the significant findings of this study. Oft cited research by Waters (1999) seems to suggest that, as a way of gaining more favorable treatment for themselves, foreign born blacks emphasize their foreign birth, their differences from the native born blacks. However, as these tables seem to demonstrate, the differences in the foreign born African ancestry students' perceptions of graduate school from those of native born African ancestry students are not great; perhaps the foreign born student' strategy of emphasizing their differences is not having much effect in graduate school.

Another possibility is that the African heritage students, whose perceptions of graduate school are much less favorable than whites and Asians, are much less prepared for graduate schoolwork than their white and Asian counterparts. Regardless, we see that for Africans to emerge with the skills, knowledge and collaborators that spell success in their careers in science we need to retool graduate schools or do a much better job at the undergraduate level of readying the African ancestry and Hispanic heritage students for the intensive work they will be doing at the graduate schools if they choose to go further. I favor the latter but this may mean fundamental changes in how courses are taught at the undergraduate level in their colleges and universities.

Overall Satisfaction and Encouragement

Tables 2.1a and 2.1b present global measures of satisfaction and encouragement to pursue the doctorate. I shall discuss Table 2.1a first, (a table I also ran for the previous book but which was not statistically significant with the previous sample).

Table 2.1a: Overall Satisfaction with Graduate School by Student Background

Student Background	Overall Satisfaction	Total
White Male	3.92	292
White Male: CO	4.31	51
White Male: NYC	4.39	110
White Male: NC	4.35	126
Asian Male	3.90	153
Black Male	3.93	28
Black Male: Native	4.24	17
Black Male: Foreign	3.45	11
Hispanic Male	3.99	100
Female	3.86	515
Total		1,088

Kruskall-Wallis equality-of-populations rank test: Chi-square with ties = 3.805, p=.433, df=4

Two aspects of Table 2.1a stand out. First, only a minority of students of any ethnic group or gender feels that they are generally encouraged to pursue the doctorate. Just under half (49.8 percent) of white males and Asian males (48.1 percent) felt generally encouraged. The results were even worse for African ancestry males (31.3%) and Hispanic heritage males (33.6 percent).

Table 2.1b: Encouraged to Pursue Doctorate by Student Background

Student Background	Yes	No	Total
White Male	49.8 (149)	50.2 (150)	299
White Male: CO	43.4 (23)	56.6 (30)	53
White Male: NYC	47.4 (54)	52.6 (60)	114
White Male: NC	54.2 (71)	45.8 (60)	131
Asian Male	48.1 (76)	51.9 (82)	158
Black Male	31.3 (10)	68.8 (22)	32
Black Male: Native	31.6 (6)	68.4 (13)	19
Black Male: Foreign	30.8 (4)	69.2 (9)	13
Hispanic Male	33.6 (39)	66.4 (77)	116
Female	43.5 (237)	56.5 (308)	545
Total	511	639	1150

Pearson chi-square test: Chi-square=12.335, p=.015, df=4

For whatever reason, no differences were apparent in overall satisfaction of the different groups with their graduate school experience. I am at a loss to understand this—perhaps it reflects the ambiguity of that metric. However, I shall attempt to ferret out the basis for different perceptions of encouragement to complete the doctoral degree when I take up the specific ways schools encourage students in the following tables in this chapter. Here, however, I shall merely note is a sobering finding, suggesting that the weed out system is quite harsh, regardless of the ethnic or gender background of the students.

The second important point is that while nearly half of the white males and Asian males feel generally encouraged to pursue the doctorate, among African ancestry and Hispanic heritage males a much smaller proportion feel generally encouraged—a third or slightly less. Women of all groups together rank roughly midway between the white males and the African ancestry males in their perceptions of feeling encouraged.

The second global measure I used was general satisfaction. Despite its ambiguity, reported overall satisfaction can suggest if the graduate school has been meeting student needs and expectations. Moreover, level of satisfaction seems to affect how highly the students rate the specific forms of encouragement they have received. (In Crippled at the Starting Gate I checked this hypothesis in my data and found that except for the dimension of "assistantships do not have burdensome responsibilities for undergraduate teaching," each of the kinds of encouragement the student receives was directly affected by overall satisfaction, i.e., the relationship between specific forms of encouragement and satisfaction was one of two-way causation).

Large differences between either the genders or ethnic groups in satisfaction, especially when they are accompanied by differences in the students' sense of being encouraged to study for the doctorate I believed may be a clue that the school's planning and implementation of services was inadequate in regard to making students of a particular background feel welcome. Consequently, I asked the survey respondents to "Please rate your overall graduate school experience at this school."

For Crippled at the Starting Gate, I had created two new measures of "general encouragement:" (1) an additive index and (2) factor analysis measure in order to better understand the differences between men and women I continued this practice in the present monograph as well.

Both indices suggested white men and Asian men are more likely to feel encouraged to pursue the doctorate than African ancestry men or Hispanic heritage men. Women again ranked somewhat between the white men and the African ancestry men but much closer to the white men than to the African ancestry men. The factor analysis scores for African ancestry and Hispanic heritage men and all women suggest, in fact, all these latter groups disagree in varying degrees their graduate school encourages them to pursue research careers (see Table 2.2).

Table 2.2: Encouraged to Pursue Doctorate by Student Background: Additive Index and Factor Analysis Indicators

Student Background	Additive Index	Factor Analysis	Total
White Male	.6012	.2256	299
White Male: CO	.5684	.1484	53
White Male: NYC	.5822	.1509	114
White Male: NC	.6326	.3268	131
Asian Male	.5649	.1056	158
Black Male	.3711	-.5436	32
Black Male: Native	.3092	-.7436	19
Black Male: Foreign	.4615	-.2514	13
Hispanic Male	.3836	-.5028	116
Female	.5289	-.0170	545
Total			1150
Kruskall-Wallis equality-of-populations rank test	Chi-square with ties = 46.556, p<.000, df=4	Chi-square with ties = 47.775, p<.000, df=4	

Specific Encouragement to Pursue Doctoral Research

When I looked at specific ways in which students could be encouraged to pursue the doctorate, the data showed important differences between white men and Asian men on one hand and African ancestry and Hispanic heritage graduate students on the other suggesting the former clearly enjoyed superior conditions than the latter groups.

Table 2.3 (below) shows perceptions in regard to whether the "school brings in speakers to talk about job opportunities in the field." Two facts immediately stand out. First, only a minority of any group asserts the school makes an effort to do this. Second, the white male students are much more likely (48.8 percent) than African ancestry male students (21.9 percent) or Hispanic heritage male students (26.7 percent) to report this form of encouragement. In these economically challenging times, it seems plausible that the graduate schools would try to motivate their students by pointing out why successful completion of the doctoral studies can pay off in good jobs in industry, academia, and government. Yet, it appears that overall the graduate schools do not make a determined effort in this regard.

Table 2.3: School Brings in Speakers to Talk About Job Opportunities by Student Background

Student Background	Yes	No	Total
White Male	48.8 (146)	51.2 (153)	299
White Male: CO	34.0 (18)	66.0 (35)	53
White Male: NYC	50.0 (57)	50.0 (57)	114
White Male: NC	54.2 (71)	54.2 (71)	131
Asian Male	41.1 (65)	58.9 (93)	158
Black Male	21.9 (7)	78.1 (25)	32
Black Male: Native	10.5 (2)	89.5 (17)	19
Black Male: Foreign	38.5 (5)	61.5 (8)	13
Hispanic Male	26.7 (31)	73.3 (85)	116
Female	41.7 (227)	58.3 (318)	545
Total	476	674	1150

Pearson chi-square test: Chi-square=22.149, p<.001, df=4

Table 2.4 shows another way in which graduate schools discourage students from successful completion of their doctoral studies. The first thing that jumps out is that substantial majorities of all graduate students claim that the teaching assistantships are imposing a heavy burden on them. The second point is even more perturbing. Compared to the students queried previously—42 percent of white men and Asian men students thought teaching assistantships did not impose a heavy burden for undergraduate education—just 28 percent of African ancestry and Hispanic heritage men felt the same way. Women were closer in this regard to white and Asian men (38.5 percent).

Table 2.4: Assistantships Do Not Require Heavy Responsibilities for Teaching Undergraduates by Student Background

Student Background	Agree	Disagree	Total
White Male	42.1 (126)	57.9 (173)	299
White Male: CO	37.7 (20)	62.3 (33)	53
White Male: NYC	53.5 (61)	46.5 (53)	114
White Male: NC	40.5 (53)	59.5 (78)	131
Asian Male	42.4 (67)	57.6 (91)	158
Black Male	28.1 (9)	71.9 (23)	32
Black Male: Native	26.3 (5)	73.7 (14)	19
Black Male: Foreign	30.8 (4)	69.2 (9)	13
Hispanic Male	27.6 (32)	72.4 (84)	116
Female	38.5 (210)	61.5 (335)	545
Total	444	706	1150

Pearson chi-square test: Chi-square=9.965, p=.041, df=4

In CATSG, my previous monograph focused on women and Asian students, I urged that teaching assistantships be dropped in favor of research assistantships. Teaching assistantships interfere with the graduate students' mastery of the subject matter and also foist on undergraduates a resentful and unmotivated group of teachers and teacher assistants that drive those undergraduate students away from the science and engineering disciplines. If we face a looming shortage of scientists and engineers with advanced credentials does it make sense to drive students away? The schools argue that they need to focus on the most promising because resources are scarce. How do they know they are getting the most promising students by following these educationally harmful practices? Perhaps they are only getting the students who are willing to tolerate the indifferent educational environment and still feel they want to continue pursuing a science or engineering degree.

The seriousness of the problem of burdensome undergraduate teaching responsibilities is best understood in the context of the students' perceptions about the adequacy of financial aid they receive for their studies. Table 2.5 (see below) shows that whereas about 85 percent of the white males report the school provides financial aid just 53 percent of African ancestry students and 60 percent of Hispanic heritage students report this assistance. I was not able to explore this issue in depth but my guess is that the African ancestry and Hispanic heritage students come from households with lower income levels on average than the white male students; therefore even if the aid package offered white male and African ancestry male and Hispanic heritage male students is equivalent in dollar terms the African ancestry and Hispanic heritage students find it harder to make ends meet because their own resources are so much less than those of the white male students. Overall, women students while less likely than white male students to report the school provides financial aid were still more likely to report this assistance from the school than were the African ancestry and Hispanic heritage students.

Table 2.5: School Provides Financial Aid by Student Background

Student Background	Yes	No	Total
White Male	84.6 (253)	15.4 (46)	299
White Male: CO	92.5 (49)	7.5 (4)	53
White Male: NYC	77.2 (88)	22.8 (26)	114
White Male: NC	87.8 (115)	12.2 (16)	131
Asian Male	79.7 (126)	20.3 (32)	158
Black Male	53.1 (17)	46.9 (15)	32
Black Male: Native	47.4 (9)	52.6 (10)	19
Black Male: Foreign	38.5 (5)	61.5 (8)	13
Hispanic Male	60.3 (70)	39.7 (46)	116
Female	75.8 (413)	24.2 (132)	545
Total	879	271	1150

Pearson chi-square test: Chi-square=38.527, $p<.001$, df=4

The data in Table 2.4-2.5 have unsettling implications for the future careers of women and minority men scientists and engineers Women professors according to Xie and Schauman (2003; also see Valian, 1998 and Fisher, 2007 p. 46), have borne a heavier burden for teaching in the STEM disciplines than men, perhaps partly accounting for disparities in publications rates. Differences in this regard have been found between major research universities and schools of lesser renown where professors carry a heavier teaching load. It appears from my survey data this differential responsibility for teaching in the STEM fields starts to fall on women and minority men during their graduate school years.

According to Table 2.6 below, a majority of white men (59 percent) and Asian men (58 percent also asserted, "informally professors encourage students to continue or start research" compared to Hispanic heritage (32 percent) and African ancestry (37.5 percent) groups. An interesting difference appears between native-born and foreign-born African ancestry students in these data. Foreign-born male students of African ancestry are twice as likely (54 percent) as native-born men students (26 percent) to report informal encouragement to start or continue research. In this regard, perhaps, Maranda's claim that foreign born students are better treated than native born African ancestry students seems to be supported by the data. However, this finding, which is based on very few cases, needs to be replicated with a much large group of cases.

Table 2.6: Encouraged to Continue or Start Research by Student Background

Student Background	Yes	No	Total
White Male	58.9 (176)	41.1 (123)	299
White Male: CO	52.8 (28)	47.2 (25)	53
White Male: NYC	57.0 (65)	43.0 (49)	114
White Male: NC	63.4 (83)	36.6 (48)	131
Asian Male	57.6 (91)	42.4 (67)	158
Black Male	37.5 (12)	62.5 (20)	32
Black Male: Native	26.3 (5)	73.7 (14)	19
Black Male: Foreign	53.8 (7)	46.2 (6)	13
Hispanic Male	31.9 (370)	68.1 (79)	116
Female	54.3 (296)	45.7 (249)	545
Total	612	538	1150

Pearson chi-square test: Chi-square=29.661, $p<.001$, df=4

Table 2.7 shows that white men (68 percent) are more likely to rate their schools highly for hiring good professors compared to 59.5 percent of Asian male students, 57 percent of women students and just 40 percent of Hispanic heritage and African ancestry men students.

represented minorities than currently if we are to meet our STEM personnel needs in the future (2010, p. 42)."

However, while I did find differences among the disciplines in that earlier study, my analysis was handicapped by the limited number of under-represented minority students I had available to me. In this present monograph, I was able to gain a better idea of the degree to which graduate schools welcome students of African ancestry and Hispanic heritage although again I was unable to do fine analysis of individual disciplines. As in the earlier study, I had to settle for examining this issue somewhat crudely, believing some data on the subject were appropriate even if they did no more than raise questions to be answered in the future.

I approached the problem from two directions. On one hand, since I did not have enough data to examine every scientific discipline separately for how it was meeting the needs of the graduate student population, I combined disciplines into two categories: "hard" and "soft sciences." The hard sciences I defined arbitrarily as physics, chemistry, engineering/computer sciences, and mathematics/statistics. Biology, behavioral sciences, and most everything else were lumped together as "soft sciences." This permitted me to look at how women and Asian men fared in each group of sciences compared to white men. Unfortunately, of course, I could not tell in which discipline within these categories the differences were generated.

I also decided to redo my earlier cross tabulations in which I looked at all men versus all women in physics, mathematics, engineering/computer science, and chemistry; I reasoned that the presence of African ancestry and Hispanic heritage men among the men would reduce the differences between the two genders. Now, however, I also looked at the following within discipline differences: white men versus all women versus Asian men versus African and Hispanic men. I was particularly interested in examining the fields of engineering and physics because women's penetration of the top ranks of those fields had been much lower than even other STEM professions (see Fisher, 2007).

Table 2.10a shows that overall satisfaction felt by hard sciences students from different component populations of the graduate school student population do not differ significantly. But if overall satisfaction does not differ among the groups, the level of encouragement does as shown in Table 2.10b.

When students were asked about whether they felt overall that the school encouraged them to complete their degrees, white male students were more likely than African ancestry and Hispanic heritage students to answer in the affirmative. Similarly, when men pursuing degrees in the hard sciences are compared on both summary measures of encouragement—the additive index of encouragement and the factor analysis measure of encouragement—white men felt more encouraged to complete their degrees than did African ancestry and Hispanic heritage students did. The results with women added in on the summary measures told the same story—white men are more encouraged than other groups although the differences between white men and Asian men on the additive index and factor analysis are not large. See Table 2.10b below).

Table 2.10a: Overall Satisfaction with Graduate School by Student Background for Hard Sciences Students

Student Background	Overall Satisfaction	Total
White Male	3.84	216
Asian Male	3.86	135
Black Male	3.92	24
Hispanic Male	3.99	70
Female	3.90	280
Total		725
Kruskall-Wallis equality-of-populations rank test	Chi-square with ties = 2.061, p=.725, df=4	

Table 2.10b: Encouraged to Pursue Doctorate by Student Background: Additive Index and Factor Analysis Indicators for Hard Sciences Students

Student Background	Additive Index	Factor Analysis	Total
White Male	.5907	.1931	222
Asian Male	.5670	.1116	140
Black Male	.4038	-.4316	26
Hispanic Male	.3904	-.4809	81
Female	.5196	-.0455	300
Total			769
Kruskall-Wallis equality-of-populations rank test	Chi-square with ties = 29.183, p<.000, df=4	Chi-square with ties = 30.402, p<.000, df=4	

Table 2.11: Assistantships Do Not Require Heavy Responsibilities for Teaching Undergraduates by Student Background for Hard Sciences Students

Student Background	Agree	Disagree	Total
White Male	43.2 (96)	56.8 (126)	222
Asian Male	42.9 (60)	57.1 (80)	140
Black Male	30.8 (8)	69.2 (18)	26
Hispanic Male	27.2 (22)	72.8 (59)	81
Female	35.7 (107)	64.3 (193)	300
Total	293	476	769

Pearson chi-square test: Chi-square=9.289, p=.054, df=4, Cramer's V=.110

Table 2.11 shows that all students in the hard sciences disciplines regardless of gender or ethnicity find teaching assistantships burdensome because they impose heavy undergraduate teaching responsibilities. However, the evidence suggests that male Hispanic heritage and African ancestry students find these responsibilities more of a burden than do white male students. The differences in percentage terms are moderate and in my opinion it is only because of the limited numbers of Hispanic heritage and African ancestry students in the study population that the statistical significance level is marginal.

Table 2.12 shows that among hard science students, white and Asian males, and even female students, feel more encouraged continuing or starting research than African ancestry male students and especially more than Hispanic heritage male students. The large difference between the African ancestry students and the Hispanic heritage male students, who report feeling encouraged in this regard at a rate about half that of white students, is one of the more surprising findings of this study.

Table 2.12: Encouraged to Continue or Start Research by Student Background for Hard Sciences Students

Student Background	Yes	No	Total
White Male	55.9 (124)	44.1 (98)	222
Asian Male	59.3 (83)	40.7 (57)	140
Black Male	42.3 (11)	57.7 (15)	26
Hispanic Male	27.2 (22)	72.8 (59)	81
Female	53.0 (159)	47.0 (141)	300
Total	399	370	769

Pearson chi-square test: Chi-square=25.413, p<.000, df=4

Among hard sciences students, the white male students (68%), far more than the African ancestry (38.5%) and Hispanic heritage (43.2%) students, thought that their schools hired good professors (table not shown). White male students, I maintain, feel encouraged by the quality of their professors and view them as examples of successful disciplinarians worthy of emulation. That fewer of the African ancestry and Hispanic heritage students feel this way may have an adverse impact on their willingness to press ahead to finish their graduate programs.

Table 2.13 shows no statistically significant differences among hard sciences students in their perceptions of access to latest equipment. Considering that the feminist scholars had focused on the hard sciences as especially resistive to women's success, at least in this regard the graduate schools were not impeding the advancement of their minority and women students. Yet, as I showed earlier (Table 2.8a) white students generally were more positive about their access to this resource—suggesting that perhaps the problem of lack of access is concentrated in the soft sciences. I shall inquire into that possibility below.

Table 2.13: School Provides Access to Latest Equipment by Student Background for Hard Sciences Students

Student Background	Yes	No	Total
White Male	47.3 (105)	52.7 (117)	222
Asian Male	47.9 (67)	52.1 (73)	140
Black Male	34.6 (9)	65.4 (17)	26
Hispanic Male	55.6 (45)	44.4 (36)	81
Female	42.3 (127)	57.7 (173)	300
Total	344	425	769

Pearson chi-square test: Chi-square=2.921, p=.571, df=4

White male students in the hard sciences according to Table 2.14 report twice as often as African ancestry that their school brings in speakers to talk about job opportunities. And compared to Hispanic heritage male students they are also considerably more likely to claim that this occurs. Still, it is interesting that in no case did a majority of hard sciences students report this occurs. Given how demanding the training for the doctoral degree in the hard sciences can be, this is surprising as it seems obvious that knowing there will be good opportunities to have a successful career would be helpful in maintaining one's commitment to the program.

Table 2.14: School Brings in Speakers to Talk About Job Opportunities by Student Background for Hard Sciences Students

Student Background	Yes	No	Total
White Male	46.4 (103)	53.6 (119)	222
Asian Male	41.4 (58)	58.6 (82)	140
Black Male	23.1 (6)	76.9 (20)	26
Hispanic Male	28.4 (23)	71.6 (58)	81
Female	40.0 (120)	60.0 (180)	300
Total	310	459	769

Pearson chi-square test: Chi-square=11.491, p=.022, df=4

White male students overwhelmingly (85 percent) claim that their school provides financial aid whereas lesser majorities of Hispanic male students (61.7 percent) make this claim (see Table 2.15). Among African ancestry students only a bare majority (53.8 percent) made his claim. Given that financial support is crucial for students embarked on this time consuming and intellectually demanding program, the comparatively lesser support for Hispanic heritage and

African ancestry students is a powerful disincentive to complete their studies successfully.

Table 2.15: School Provides Financial Aid by Student Background for Hard Sciences Students

Student Background	Yes	No	Total
White Male	85.1 (189)	14.9 (33)	222
Asian Male	77.9 (109)	22.1 (31)	140
Black Male	53.8 (14)	46.2 (12)	26
Hispanic Male	61.7 (50)	38.3 (31)	81
Female	74.0 (222)	26.0 (78)	300
Total	584	185	769

Pearson chi-square test: Chi-square=27.075, p<.000, df=4

More than three-quarters (77.9 percent) of white male students in the hard sciences report that the school provides opportunities for them to work on research problems compared to less than half of the Hispanic heritage students (48.1 percent) and 61.5 percent of the African ancestry students as shown in Table 2.16.

Table 2.16: School Provides Opportunities to Work on Research Problems for Hard Sciences Students

Student Background	Yes	No	Total
White Male	77.9 (173)	22.1 (49)	222
Asian Male	77.1 (108)	22.9 (32)	140
Black Male	61.5 (16)	38.5 (10)	26
Hispanic Male	48.1 (39)	51.9 (42)	81
Female	68.3 (205)	31.7 (95)	300
Total	541	228	769

Pearson chi-square test: Chi-square=29.904, p<.000, df=4

African ancestry males (3.73 out of 5) were much less likely to feel that professors were respectful of ethnic differences than were white male students (4.37 out of 5) according to Table 2.17. However, surprisingly, Hispanic male students were more likely (4.54 out of 5) to feel the professors were respectful of these differences than white male students.

Table 2.17: Professors Respectful of Gender and Ethnic Differences by Student Background for Hard Sciences Students

Student Background	Respectful of: Gender Differences	Ethnic Differences
White Male	4.31 (207)	4.37 (211)
Asian Male	4.29 (136)	4.28 (137)
Black Male	4.20 (25)	3.73 (26)
Hispanic Male	4.50 (80)	4.54 (81)
Female	4.13 (289)	4.12 (287)
Total	737	742
Kruskall-Wallis equality-of-populations rank test	Chi-square with ties = 15.292, p=.004, df=4	Chi-square with ties = 29.612, p<.000, df=4

SOFT SCIENCES

Whereas social scientists have written extensively about the resistance in hard sciences to penetration of the higher ranks by women (and, by implication, by African ancestry and Hispanic heritage disciplinarians of either gender), they have not been so focused on the other so-called "soft sciences." Indeed, whatever research has been done on the softer sciences such as sociology and psychology has pointed to progress women have made in gaining equality in publication productivity (see, for example Stack, 1994 and other references in V. Valian, 1998) and some measure of representation in the higher echelons of these disciplines. However, as shown in the results reported below the picture is not as rosy as might be suggested by these studies. Indeed, the possibility that further gains might be stymied—or even that hard won gains made by women and minorities over the past few decades might be erased—cannot be dismissed as simply implausible. The road to equal opportunity regardless of gender or ethnic background will be difficult and time consuming, with the best available estimates suggesting it will be decades before it occurs (see Kaminsky and Geisler).

Although overall satisfaction of women students, African ancestry students and Hispanic heritage students in the soft sciences (e.g., social sciences, biology) does not differ to a statistically significant degree on the five percent criterion (See Table 2.18), the encouragement variables tell a more unsettling tale. Table 2.19 shows (based on the two summary measures) that white male students are considerably more likely to report feeling encouraged to complete their studies successfully than either African ancestry or Hispanic heritage males.

Table 2.18: Overall Satisfaction with Graduate School by Student Background for Soft Sciences Students

Student Background	Overall Satisfaction	Total
White Male	4.12	75
Asian Male	4.12	17
Black Male	4.00	4
Hispanic Male	4.00	30
Female	3.80	233
Total		359
Kruskall-Wallis equality-of-populations rank test	Chi-square with ties = 8.373, p=.079, df=4	

Table 2.19: Encouraged to Pursue Doctorate by Student Background for Soft Sciences Students

Student Background	Yes	No	Additive Index	Factor Analysis
White Male	52.6 (40)	47.4 (36)	.6299 (76)	.3136 (76)
Asian Male	52.9 (9)	47.1 (8)	.5368 (17)	.0273 (17)
Black Male	0.0 (0)	100.0 (6)	.2292 (6)	-1.0292 (6)
Hispanic Male	37.1 (13)	62.9 (22)	.3679 (34)	-.5535 (34)
Female	43.6 (106)	56.4 (137)	.5386 (237)	.0125 (237)
Total	168	209	370	370

In Table 2.10, I showed that majorities of all ethnic component populations among male graduate students in the hard sciences, as well as female graduate students, complained that assistantships imposed burdensome undergraduate teaching responsibilities. In fact, whereas the differences among the groups in the hard sciences were marginally significant, they were not statistically significant among the groups in the soft sciences as demonstrated in Table 2.20.

The message in Table 2.20 is ominous. Now, it is clear, among soft sciences students the problem of burdensome assistantship responsibilities is at least as serious, if not more so, than in the hard sciences. Although as I stated, the differences were not statistically significant, surprisingly, women graduate students in the soft sciences (58.4 percent) were less likely than any other group in the soft sciences segment of the graduate student population to make this complaint about burdensome undergraduate teaching duties. African ancestry males (83.3 percent) and Hispanic heritage males (71.4 percent) were the most likely to complain. Keep in mind that this kind of complaint is a serious disincentive to completing the graduate school program successfully because the African ancestry and Hispanic heritage male students are likelier to have financial hardships that make them more dependent than white males on the financial support of the assistantship to continue their studies.

Table 2.20: Assistantships Do Not Require Heavy Responsibilities for Teaching Undergraduates by Student Background for Soft Sciences Students

Student Background	Agree	Disagree	Total
White Male	39.5 (30)	60.5 (46)	76
Asian Male	35.3 (6)	64.7 (11)	17
Black Male	16.7 (1)	83.3 (5)	6
Hispanic Male	28.6 (10)	71.4 (25)	35
Female	41.6 (101)	58.4 (142)	243
Total	229	148	377

Pearson chi-square test: Chi-square=3.616, p=.461, df=4

Table 2.21 shows white men students in the soft sciences claimed they were encouraged to start or continue research far more than other groups (except females who differed from white males moderately). Fully two-thirds (67.1 percent) of the white male graduate students in the soft sciences versus 41.2 of Asian male students, 42.0 percent of Hispanic male students and just 16.7 percent of the African ancestry students. The numbers of African ancestry males for whom data were available on this variable is small but the overall message of the table—that the schools are encouraging the white male students far more than others to complete their doctoral studies—is impossible to ignore.

Table 2.21: Encouraged to Continue or Start Research by Student Background for Soft Sciences Students

Student Background	Yes	No	Total
White Male	67.1 (51)	32.9 (25)	76
Asian Male	41.2 (7)	58.8 (10)	17
Black Male	16.7 (1)	83.3 (5)	6
Hispanic Male	42.9 (15)	57.1 (20)	35
Female	55.6 (135)	44.4 (108)	243
Total	209	168	377

Pearson chi-square test: Chi-square=11.482, p=.022, df=4

Table 2.22 shows that white male students (67.1 percent) to a considerably greater degree than other students in the soft sciences respondent group believe their school hires good professors. Whereas two-thirds of the white male students felt this way, only about a third (34.3 percent) of the Hispanic heritage students agreed with this statement; in the other groups, roughly half of the respondents—fifty percent of the African ancestry male students and 55 percent of the female students—felt this way.

Table 2.22: School Hires Good Professors by Student Background for Soft Sciences Students

Student Background	Yes	No	Total
White Male	67.1 (51)	32.9 (25)	76
Asian Male	52.9 (9)	47.1 (8)	17
Black Male	50.0 (3)	50.0 (3)	6
Hispanic Male	34.3 (12)	65.7 (23)	35
Female	55.1 (134)	44.9 (109)	243
Total	209	168	377

Pearson chi-square test: Chi-square=10.650, p=.031, df=4

Part of the training for real jobs that should be integral to graduate school education is giving students access to the latest equipment, the kind likely to be found in the thriving businesses and academic research centers doing cutting edge research. Table 2.23 shows that white men students (60.5 percent) in the soft sciences are almost three times as likely to report access to the latest equipment as Hispanic heritage male students (22.9 percent) and nearly twice as likely as African ancestry students (33.3 percent) to report they had access to the latest equipment in their graduate education.

Table 2.23: School Provides Access to Latest Equipment by Student Background for Soft Sciences Students

Student Background	Yes	No	Total
White Male	60.5 (46)	39.5 (30)	76
Asian Male	47.1 (8)	52.9 (9)	17
Black Male	33.3 (2)	66.7 (4)	6
Hispanic Male	22.9 (8)	77.1 (27)	35
Female	43.6 (106)	56.4 (137)	243
Total	170	207	377

Pearson chi-square test: Chi-square=14.875, p=.005, df=4

Speakers discussing opportunities in the field for recent graduates can be motivating for graduate students to succeed in finishing their studies. According to Table 2.24, white male students (55.3 percent) were more likely than any other group to report their school provides such speakers. African ancestry male students (16.7 percent) and Hispanic heritage male students (22.9 percent) were much less likely to report their school provides such speakers for them.

Table 2.24: School Brings in Speakers to Talk about Job Opportunities by Student Background for Soft Sciences Students

Student Background	Yes	No	Total
White Male	55.3 (42)	44.7 (34)	76
Asian Male	35.3 (6)	64.7 (11)	17
Black Male	16.7 (1)	83.3 (5)	6
Hispanic Male	22.9 (8)	77.1 (27)	35
Female	43.6 (106)	56.4 (137)	243
Total	163	214	377

Pearson chi-square test: Chi-square=12.579, p=.014, df=4

Pursuing the doctorate is a long term quest, perhaps seven years from start to finish, and represents a considerable financial sacrifice for students. For students from less affluent family backgrounds, likely to be true of the African ancestry and Hispanic heritage students, it would probably be an unattainable objective without financial assistance from the graduate school. Table 2.25 shows that Asian male students (94.1 percent) and white male students (82.9 percent) were much more likely to have this kind of necessary support than the African ancestry students (50 percent) or Hispanic heritage graduate students (57.1 percent) in the soft sciences.

Table 2.25: School Provides Financial Aid by Student Background for Soft Sciences Students

Student Background	Yes	No	Total
White Male	82.9 (63)	17.1 (13)	76
Asian Male	94.1 (16)	5.9 (1)	17
Black Male	50.0 (3)	50.0 (3)	6
Hispanic Male	57.1 (20)	42.9 (15)	35
Female	78.2 (190)	21.8 (53)	243
Total	292	85	377

Pearson chi-square test: Chi-square=14.925, p=.005, df=4

Opportunities to work on real research problems, perhaps as a research assistant to a senior professor on a grant, are valuable training for junior scientists aspiring to succeed in their chosen discipline. Table 2.26 shows that

78.9 percent of the white male graduate students in the soft sciences reported they were able to get this valuable experience compared 48.6 percent of the Hispanic heritage students and none (0 percent) of the African ancestry students in the sample.

Table 2.26: School Provides Opportunities to Work on Research Problems by Student Background for Soft Sciences Students

Student Background	Yes	No	Total
White Male	78.9 (60)	21.1 (16)	76
Asian Male	70.6 (12)	29.4 (5)	17
Black Male	0.0 (0)	100.0 (6)	6
Hispanic Male	48.6 (17)	51.4 (18)	35
Female	69.5 (169)	30.5 (74)	243
Total	258	119	377

Pearson chi-square test: Chi-square=23.465, p=.000, df=4

Table 2.27 shows that in the soft sciences the student groups do not differ significantly in the ratings of professors' respect for ethnic differences. However, on gender differences, marginal statistical differences were detected—women (4.08 out of 5) gave the professors the lowest scores. Interestingly, African ancestry males scored the professors highest of all the groups on this dimension while Asian male students had scores (4.12 out of 5) little different from the women.

Table 2.27: Professors Respectful of Gender and Ethnic Differences by Student Background for Soft Sciences Students

Student Background	Gender Differences	Ethnic Differences
White Male	4.24 (76)	4.33 (76)
Asian Male	4.12 (17)	4.06 (17)
Black Male	0.0 (6)	100.0 (6)
Hispanic Male	48.6 (34)	51.4 (34)
Female	69.5 (237)	30.5 (232)
Total	370	365
Kruskall-Wallis equality of proportions rank test	Chi-square with ties = 9.142 p=.058, df=4	Chi-square with ties = 6.678 p=.154, df=4

Mediating Role of Atmosphere - Intellectual, Snobbish, or Friendly

School milieu plays a role in as a mediating variable in whether the graduate students feel encouraged in specific ways to earn the doctorate in their specific fields. In the following section I look at this relationship for African ancestry and Hispanic heritage students—a group I could not discuss in my earlier book CATSG for lack of cases.

Table 2.28a shows that African ancestry students agreeing or strongly agreeing that their school had a friendly atmosphere were more likely than those not agreeing about this to say they felt the school faculty encouraged them to start or continue their research. African ancestry respondents were also slightly more likely to say that the school provides access to the latest equipment.

Table 2.28a: Encouraged to Continue or Start Research by Friendly Atmosphere for Black Students

Friendly Atmosphere	Yes	No	Total
Strongly Agree/Agree	44.1 (26)	55.9 (33)	59
All Others	20.0 (3)	80.0 (12)	15
Total	29	45	74

Pearson chi-square test: Chi-square=2.907, p=.088, df=1; Gamma=.518

Table 2.28b: School Provides Access to Latest Equipment by Friendly Atmosphere for Black Students

Friendly Atmosphere	Yes	No	Total
Strongly Agree/Agree	44.1 (26)	55.9 (33)	59
All Others	20.0 (3)	80.0 (12)	15
Total	29	45	74

Pearson chi-square test: Chi-square=2.907, p=.088, df=1; Gamma=.518

Table 2.28c shows that African ancestry students strongly agreeing or agreeing the school was friendly were no more likely than the same group in less friendly schools to claim that the school did not impose heavy responsibilities in its assistantships for teaching undergraduates.

Table 2.28c: Assistantships Do Not Require Heavy Responsibilities for Teaching Undergraduates by Friendly Atmosphere for Black Students

Friendly Atmosphere	Yes	No	Total
Strongly Agree/Agree	30.5 (18)	69.5 (41)	59
All Others	20.0 (3)	80.0 (12)	15
Total	21	53	74

Pearson chi-square test: Chi-square=.650, p=.420, df=1

Table 2.29 shows that among those African ancestry students agreeing or strongly agreeing that the school has an intellectual atmosphere, assistantships were less likely to be seen as imposing a heavy burden of undergraduate educational responsibility (although the table just misses statistical significance at the .1 level).

Table 2.29: Assistantships Do Not Require Heavy Responsibilities for Teaching Undergraduates by Intellectual Atmosphere for Black Students

Intellectual Atmosphere	Agree	Disagree	Total
Strongly Agree/Agree	30.9 (21)	69.1 (47)	68
All Others	0.0 (0)	100.0 (6)	6
Total	21	53	74

Pearson chi-square test: Chi-square=2.587, p=.108, df=1; Cramer's V=.187

Among male graduate students of African ancestry, however, intellectual atmosphere was negatively related to school provides access to latest equipment although the difference was not great. (See Table 2.30a). Similarly, male graduate students of African ancestry quite strongly disagreed that the school hired good professors when they felt the atmosphere was snobbish (Table 2.30b). In this regard, the graduate students at Ivy League institutions and other elite private institutions such as Duke and NYU were more likely to claim their schools were snobbish according to my data than students at other graduate schools (see Table 5.18, p. 119 in CATSG).

Table 2.30a: School Provides Access to Latest Equipment by Intellectual Atmosphere for Black Male Students

Intellectual Atmosphere	Yes	No	Total
Strongly Agree/Agree	36.2 (34)	63.8 (60)	94
All Others	45.5 (481)	54.5 (575)	1056
Total	515	635	1150

Pearson chi-square test: Chi-square=3.071, p=.080, df=1

Table 2.30b: School Hires Good Professors by Snobbish Atmosphere for Black Male Students

Snobbish Atmosphere	Yes	No	Total
Strongly Agree/Agree	0.0 (0)	100.0 (5)	5
All Others	48.1 (13)	51.9 (14)	27
Total	13	19	32

Pearson chi-square test: Chi-square=4.055, p=.044, df=1

Similar to all graduate students of African ancestry, female graduate students of African ancestry agreeing or strongly agreeing that their school had a friendly atmosphere, were much more likely than those not agreeing about this to say the school provides (a) financial aid and (b) access to the latest equipment (See Tables 2.31a-b below) while those who felt their school had an intellectual atmosphere were much more likely than those not agreeing about this to assert their school provides financial aid (Table 2.32).

Table 2.31a: School Provides Financial Aid by Friendly Atmosphere for Black Female Students

Friendly Atmosphere	Yes	No	Total
Strongly Agree/Agree	56.3 (18)	43.7 (14)	32
All Others	20.0 (2)	80.0 (8)	10
Total	20	22	42

Pearson chi-square test: Chi-square=4.014, p=.045, df=1

Table 2.31b: School Provides Access to Latest Equipment by Friendly Atmosphere for Black Female Students

Friendly Atmosphere	Yes	No	Total
Strongly Agree/Agree	50.0 (16)	50.0 (16)	32
All Others	20.0 (2)	80.0 (8)	10
Total	18	24	42

Pearson chi-square test: Chi-square=2.800, p=.094, df=1; Gamma=.600

Table 2.32: School Provides Financial Aid by Intellectual Atmosphere for Black Female Students

Intellectual Atmosphere	Yes	No	Total
Strongly Agree/Agree	51.3 (20)	48.7 (19)	39
All Others	0.0 (0)	100.0 (3)	3
Total	20	22	42

Pearson chi-square test: Chi-square=2.937, p=.087, df=1; Cramer's V=.264

Table 2.33 shows that among Hispanic heritage graduate students significantly more at schools with friendly atmospheres agreed or strongly agreed that "assistantships do not require heavy responsibilities for teaching undergraduates" compared with schools whose atmospheres were less friendly. In other words, the pain of undergraduate teaching burdens was ameliorated to some degree for the Hispanic students at schools with friendly atmospheres.

Table 2.33: Assistantships Do Not Require Heavy Responsibilities for Teaching Undergraduates by Friendly Atmosphere for Hispanic Students

Friendly Atmosphere	Agree	Disagree	Total
Strongly Agree/Agree	29.2 (52)	70.8 (126)	178
All Others	7.1 (2)	92.9 (26)	28
Total	54	152	206

Pearson chi-square test: Chi-square=6.093, p=.014, df=1

Hispanic heritage students at schools rated as having an "intellectual atmosphere" were significantly more likely than those at schools not having such an atmosphere to agree or strongly agree that the school brings in speakers to talk about job opportunities (see Table 2.34a). Also at schools with an intellectual atmosphere, according to Table 2.34b, it is possible that Hispanic heritage students were more likely to see the school as providing access to the latest equipment for research compared to schools lacking an intellectual atmosphere. However, this table narrowly missed statistical significance at the .10 criterion level.

Table 2.34a: School Brings in Speakers to Talk About Job Opportunities by Intellectual Atmosphere for Hispanic Students

Intellectual Atmosphere	Yes	No	Total
Strongly Agree/Agree	27.7 (52)	72.3 (136)	188
All Others	0.0 (0)	100.0 (18)	18
Total	52	154	206

Pearson chi-square test: Chi-square=6.660, p=.010, df=1

Table 2.34b: School Provides Access to Latest Equipment by Intellectual Atmosphere for Hispanic Students

Intellectual Atmosphere	Yes	No	Total
Strongly Agree/Agree	35.6 (67)	64.4 (121)	188
All Others	16.7 (3)	83.3 (15)	18
Total	70	136	206

Pearson chi-square test: Chi-square=2.636, p=.104, df=1; Gamma=.469

At schools with a snobbish atmosphere, Hispanic heritage students were significantly more likely to feel that assistantships were imposing heavy responsibilities for educating undergraduates than at schools not having a snobbish atmosphere.

Table 2.35: Assistantships Do Not Require Heavy Responsibilities for Teaching Undergraduates by Snobbish Atmosphere for Hispanic Students

Snobbish Atmosphere	Agree	Disagree	Total
Strongly Agree/Agree	14.0 (6)	86.0 (37)	43
All Others	29.4 (48)	70.6 (115)	163
Total	54	152	206

Pearson chi-square test: Chi-square=4.223, p=.040, df=1

Male Hispanic heritage students at schools with a friendly atmosphere report that that the school provides opportunities to work on research problems to the graduate students significantly more often than their counterparts at less friendly schools. Similarly, at friendly schools they are more likely to assert "assistantships do not require heavy responsibilities for teaching undergraduates" (see Tables 2.36a-b).

Table 2.36a: School Provides Opportunities to Work on Research Problems by Friendly Atmosphere for Hispanic Male Students

Friendly Atmosphere	Yes	No	Total
Strongly Agree/Agree	52.0 (53)	48.0 (49)	102
All Others	21.4 (3)	78.6 (11)	14
Total	56	60	116

Pearson chi-square test: Chi-square=4.596, p=.032, df=1

Table 2.36b: Assistantships Do Not Require Heavy Responsibilities for Teaching Undergraduates by Friendly Atmosphere for Hispanic Male Students

Friendly Atmosphere	Agree	Disagree	Total
Strongly Agree/Agree	30.4 (31)	69.6 (71)	102
All Others	7.1 (1)	92.9 (13)	14
Total	32	84	116

Pearson chi-square test: Chi-square=3.331, p=.068, df=1; Gamma=.700

An intellectual atmosphere was likely to have some effect on Hispanic heritage male students' feeling generally encouraged to pursue the doctoral degree and also with their perceiving that the school brings in speakers to talk about job opportunities in the field.

Table 2.37a: Encouraged to Pursue Doctorate by Intellectual Atmosphere for Hispanic Male Students

Intellectual Atmosphere	Yes	No	Total
Strongly Agree/Agree	36.2 (38)	63.8 (67)	105
All Others	9.1 (1)	90.9 (10)	11
Total	39	77	116

Pearson chi-square test: Chi-square=3.276, p=.070, df=1; Gamma=.700

Table 2.37b: School Brings in Speakers to Talk About Job Opportunities by Intellectual Atmosphere for Hispanic Male Students

Intellectual Atmosphere	Yes	No	Total
Strongly Agree/Agree	29.5 (31)	70.5 (74)	105
All Others	0.0 (0)	100.0 (11)	11
Total	31	85	116

Pearson chi-square test: Chi-square=4.432, p=.035, df=1

A snobbish atmosphere, however, had some negative effects on Hispanic heritage male students' perceptions of the welcome they were receiving. Hispanic heritage male students who saw their school as having a snobbish atmosphere were less likely to claim the school hires good professors than others who did not see the school as snobbish. They also were more likely to feel the school imposes heavy undergraduate teaching responsibilities on graduate students receiving assistantships.

Table 2.38a: School Hires Good Professors by Snobbish Atmosphere for Hispanic Male Students

Snobbish Atmosphere	Yes	No	Total
Strongly Agree/Agree	29.8 (28)	70.2 (66)	94
All Others	39.4 (416)	60.6 (640)	1056
Total	444	706	1150

Pearson chi-square test: Chi-square=3.361, p=.067, df=1

Table 2.38b: Assistantships Do Not Require Heavy Responsibilities for Teaching Undergraduates by Snobbish Atmosphere for Hispanic Male Students

Snobbish Atmosphere	Agree	Disagree	Total
Strongly Agree/Agree	8.7 (2)	91.3 (21)	23
All Others	32.3 (30)	67.7 (63)	93
Total	32	84	116

Pearson chi-square test: Chi-square=5.125, p=.024, df=1

Hispanic heritage women students were less likely to regard assistantships as imposing heavy undergraduate teaching responsibilities if they saw the campus as having a friendly atmosphere and more likely to claim the school provides access to the latest equipment if they saw the school as having an intellectual atmosphere.

Table 2.39: Assistantships Do Not Require Heavy Responsibilities for Teaching Undergraduates by Friendly Atmosphere for Hispanic Female Students

Friendly Atmosphere	Agree	Disagree	Total
Strongly Agree/Agree	27.6 (21)	72.4 (55)	76
All Others	7.1 (1)	92.9 (13)	14
Total	22	68	90

Pearson chi-square test: Chi-square=2.687, p=.101, df=1; Gamma=.665

Table 2.40: School Provides Access to Latest Equipment by Intellectual Atmosphere for Hispanic Female Students

Intellectual Atmosphere	Yes	No	Total
Strongly Agree/Agree	31.3 (26)	68.7 (57)	83
All Others	0.0 (0)	100.0 (7)	7
Total	26	64	90

Pearson chi-square test: Chi-square=3.084, p=.079, df=1; Cramer's V=.185

SEXISM OF MALE GRADUATE STUDENTS: IMPLICATIONS FOR A DIVERSIFIED STUDENT BODY IN SCIENCE AND ENGINEERING

Up to now I have focused on encouragement to succeed in graduate school, and I have emphasized that white male students feel more encouraged to succeed in their graduate school studies than other component groups of the graduate student body, especially women and such minorities as African ancestry and Hispanic heritage students. Greater positive encouragement for white male students to succeed, however, is not the only way in which graduate schools perpetuate the gender gap in science and engineering. Practices and attitudes that effectively discourage others also are part of the picture.

The main practice that discourages women, I maintain, is that their financial assistance, while adequate, is often in the form of teaching assistantships and, as I reported in CATSG (see especially Table 2.3, p.35 and accompanying text), women report that these assistantships impose heavy responsibilities for teaching undergraduates. However, this is not the only hurdle to success in graduate school that women and minorities face. Another problem is the sexist attitudes of fellow students, the belief that, for example, men make better professors, better scientists, and better team members.

In this section, I shall present evidence of male sexism. Because in my previous study CATSG, proportionately far more Asian men seemed to exhibit sexist attitudes than other groups, I will start with the Asian male graduate students

Table 2.41 shows that Asian men in the "hard sciences" (22.9 percent) were more likely than any other group of graduate students to agree that men make better professors. In fact, they were the only group in the hard sciences where more than five percent of the men agreed with this stimulus.

Table 2.41: Men Make Better Professors by Student Background for Hard Sciences Students

Student Background	Strongly Agree/Agree	All Others	Total
White Male	3.2 (7)	96.8 (215)	222
Asian Male	22.9 (32)	77.1 (108)	140
Black Male	0.0 (0)	100.0 (26)	26
Hispanic Male	4.9 (4)	95.1 (77)	81
Female	7.3 (22)	92.7 (278)	300
Total	65	704	769

Pearson chi-square test: Chi-square=49.776, p<.000, df=4

Table 2.42 shows that Asian men in the "hard sciences" (27.9 percent) were more likely than any other group of graduate students to agree that men make better scientists and engineers. However, this time, 8.6 percent of the white males also agreed.

Table 2.42: Men Make Better Scientists and Engineers by Student Background for Hard Sciences Students

Student Background	Strongly Agree/Agree	All Others	Total
White Male	8.6 (19)	91.4 (203)	222
Asian Male	27.9 (39)	72.1 (101)	140
Black Male	3.8 (1)	96.2 (25)	26
Hispanic Male	6.2 (5)	93.8 (76)	81
Female	6.0 (18)	94.0 (282)	300
Total	82	687	769

Pearson chi-square test: Chi-square=54.310, p<.000, df=4

Table 2.43 shows that Asian males in the "hard sciences" (14.3 percent) were more likely than any other group of graduate students to agree that men make better team members. This time 7.4 percent of the Hispanic males also agreed with this stimulus.

Table 2.43: Men Make Better Team Members by Student Background for Hard Sciences Students

Student Background	Strongly Agree/Agree	All Others	Total
White Male	5.0 (11)	95.0 (211)	222
Asian Male	14.3 (20)	85.7 (120)	140
Black Male	3.8 (1)	96.2 (25)	26
Hispanic Male	7.4 (6)	92.6 (75)	81
Female	6.3 (19)	93.7 (281)	300
Total	57	712	769

Pearson chi-square test: Chi-square=12.582, p=.014, df=4

Based on the results from CATSG, e.g., I expected that Asian male students in the hard sciences were more likely to harbor sexist views than other ethnic groups. And, of course, since women were having the most difficult time in the "hard sciences" breaking into the top echelons of the discipline, a point emphasized in the writings of feminist scholars such as Wajcman, it also came as no great surprise to find that sexism was a serious problem among hard sciences students, in particular, the Asian students who are prominent in the student body within these disciplines.

In contrast to the hard sciences where I was confident I would find significant student sexism, I was less sure of what I would encounter in the soft sciences. I was pleased, therefore, to find that when I asked the soft sciences students if they agreed that men make better professors I did not find statistically significant differences (.05 criterion) between Asian and other student cohorts. In fact, among the soft sciences students, African ancestry males and Asian males (!) unanimously disagreed that men make better professors.

Again, when I looked at soft science students' responses to the stimulus "Men make better scientists and engineers," I found that no statistically significant differences among the groups appeared. African ancestry and Hispanic heritage males unanimously disagreed with this stimulus. However, 11.8 percent of the Asian males agreed. The actual number of cases is small (N=2) but it is somewhat perturbing nonetheless.

Finally, when I looked at soft science students responses to the stimulus "Men make better team members," again I found no statistically significant differences among the groups appeared in the data. However, 8.6 percent of the Hispanic male students agreed with the stimulus.

The next group of tables considers the impact of a climate of sexism on student attitudes about whether men make better professors, scientists and team member. Before I present the data, I need to make an explanatory comment regarding how I assessed a climate of sexism. For this purpose, I first looked at how students rated their professors on respect for gender differences or above average on this dimension on a scaled variable. I next looked at the array of the responses and created a dichotomous group membership variable by dividing the respondents into two groups. The students who rated their professors average or below I refer to as "Group A" and those who rated their professors above average I refer to as "Group B."

I then inquired into how students in these two groups feel in regard to whether men make better professors, scientists, and team members. My examination of Asian male students for example, showed that group A males: (1) felt their classmates believed men make better professors (24.2 percent) more often than group B males (14.3 percent); (2) felt their classmates believed men make better scientists (23.1 percent) about the same as group B males (22.2 percent); and also felt men make better team members (14.3) about as often as group B males (12.7 percent) (see Tables 2.44a-c below).

Table 2.44a: Classmates Think Men Make Better Professors by Ethnicity for Men who Rate Professors Average or Below on Respect for Gender Differences

Average or Below Respect for Gender Differences	Strongly Agree/Agree	All Others	Total
White Male	6.4 (10)	93.6 (147)	157
Asian Male	24.2 (22)	75.8 (69)	91
Black Male	12.5 (2)	87.5 (14)	16
Hispanic Male	10.0 (5)	90.0 (45)	50
Total	39	275	314

Pearson chi-square test: Chi-square=17.115, p=.001, df=3

Table 2.44b: Classmates Think Men Make Better Scientists and Engineers by Ethnicity for Men who Rate Professors Average or Below on Respect for Gender Differences

Average or Below Respect for Gender Differences	Strongly Agree/Agree	All Others	Total
White Male	8.3 (13)	91.7 (144)	157
Asian Male	23.1 (21)	76.9 (70)	91
Black Male	25.0 (4)	75.0 (12)	16
Hispanic Male	10.0 (5)	90.0 (45)	50
Total	43	271	314

Pearson chi-square test: Chi-square=12.980, p=.005, df=3

Table 2.44c: Classmates Think Men Make Better Team Members by Ethnicity for Men who Rate Professors Average or Below on Respect for Gender Differences

Average or Below Respect for Gender Differences	Strongly Agree/Agree	All Others	Total
White Male	5.7 (9)	94.3 (148)	157
Asian Male	14.3 (13)	85.7 (78)	91
Black Male	12.5 (2)	87.5 (14)	16
Hispanic Male	10.0 (5)	90.0 (45)	50
Total	29	285	314

Pearson chi-square test: Chi-square=5.305, p=.151, df=3

What can I say about other ethnic groups? Among group A male students, according to Table 2.44a-c, 12.5 percent of the African ancestry students and ten percent of the Hispanic heritage students saw their classmates as thinking men make better professors. The students' perceptions of sexism among classmates suggest a higher proportion of students hold these views than are suggested by the direct data of students being asked how they personally felt. For instance, none of the African male students either in group A or group B claimed they felt men make better professors. Now we see that when they are asked how their classmates feel, 12.5 percent of group A African male students assert that their classmates do feel that men make better professors—considerably more than admit they themselves feel this way.

I was encouraged to see that variations existed within ethnic groups under different atmosphere conditions in respect to professors' respect for ethnic differences or respect for gender differences. I next wanted to know if I would see variations depending on whether the students were in the hard sciences or in the soft sciences. From previous research, I expected that students who were drawn to the "hard sciences," for example, would have more sexism than students who were drawn to the soft sciences. Consequently, I wanted to look first at students by ethnicity within "hard sciences" and then within "soft sciences" to investigate this hypothesis.

The data presented suggest that sexism continues to be a serious problem within the S &E disciplines, especially among hard sciences students. However, small minorities of soft sciences students, including Hispanic heritage students as well as Asian students, also show evidence of having sexist attitudes.

SUMMARY AND CONCLUSIONS

White male students are more welcome in the graduate schools compared to other groups studied. The differences are usually not large, yet the pattern of greater encouragement given to the white male graduate students is impossible to ignore: more opportunities to do research on problems, more informal encouragement to start or continue research, more efforts to bring in speakers to talk about opportunities in the field, more financial assistance, more access to the latest equipment.

Yet changing this situation will not be simple and it definitely will not be cheap. If however, we are serious about increasing the number of Hispanic heritage and African ancestry graduate students receiving doctorates and improving their competence we must shoulder this burden. I see no other way to make conditions better for all segments of the graduate student population in the sciences and engineering.

To those who object about the cost I respond, what is the alternative? Making conditions worse for white male students would reduce the inequities but would be counterproductive. It is analogous to the humorous saying "Morale is terrible around here. We need to continue the beatings until the morale improves!"

The costs will come from three expensive steps. First, it may require expanding school faculties to provide new senior faculty slots for women and minorities. These are the most expensive faculty positions. But star faculty drawn from the ranks of women and minorities are necessary role models for women and minority students. They embody science's commitment to equal opportunity regardless of race, creed, national origin, or gender.

Can other steps be taken? Two big changes would be financial support for postgraduate education and greater diversification of the student body. Graduate schools have adopted some very bad practices—teaching assistantships and recruiting large numbers of full tuition paying students from abroad are two prime examples. Perhaps these practices did not originate because of because of tight budgets but changing them will also be costly and unlikely to happen without more money.

Recruiting large numbers of foreign students is tempting. First the foreign students frequently are superior in their academic preparation compared to American applicants. Second, the foreign students are often interested in enrolling in science and engineering programs for which few American students can be found. American International Education Foundation (AIEF, n.d.) reported in their newsletter online that "foreign graduate students have helped many U.S. universities maintain valuable research programs when enough U.S. applicants weren't available." However, the downside of this strategy is that the schools begin to reduce the quality of instruction secure in the knowledge that they will not get any political heat from angry parents or their legislators for skimping on educational quality for foreign students.

Teaching assistantships is another malicious practice. It cheats both the graduate students who need to spend their time mastering their discipline rather than doing mundane but time consuming work grading undergraduate student papers, teaching elementary courses, and the like. It cheats the undergraduates as well. The graduate assistants may be poor lecturers whose foreign accents are difficult to follow in a large lecture hall. The assistants are also not very motivated since time spent on undergraduates is time taken from their work on their graduate degrees. Yet it is difficult to fight this practice of depending on teaching assistants because in the short term, cheap labor—graduate students—saves the schools money on tuition for undergraduates. And since the less promising graduate students are usually drafted to do this labor the school can still pretend it is preparing promising students for careers in the discipline.

In my previous book Crippled at the Starting Gate I urged a "GI bill" to send more students to graduate school in the sciences and engineering. The wisdom of this investment seems so patently obvious to me I wonder why it is not already enacted. You merely need to read the financial news to see that (1) young people with technical skills are the ones who have the lowest unemployment rates among the younger people looking for work; (2) people with technical skills command the best salaries when they are hired. Furthermore, a large pool of college graduates with few marketable skills is often associated with social instability and economic distress in a society.

Our economy is distorted by spending vast sums to lock up young people for long periods of time. For example, a high proportion of young men of African ancestry, perhaps as much as a quarter of them, have spent time in jail. Does it not make more sense to spend money on educating them for productive work in an economy that is creating jobs requiring skills rather than just brawn?

Chapter Three
Women Graduate Students: Some Are More Equal Than Others

My basic hypothesis here, and in *Crippled at the Starting Crippled at the Starting Gate* (Fisher, 2010), is that "reproduction of the profession occurs in a way that selects for people with similar social, cultural, and economic characteristics to those already in the profession" (Etzkowitz et al., p. 52). Operationally, I translated this rather abstract statement into the following hypothesis: disciplines would prefer to recruit and select white males as they had done in the past.

While I was confident that graduate schools might prefer white men, I was equally certain they were not hostile to the idea of women graduate students. More precisely, although I had little reason to believe graduate schools were ecstatic about the prospect of women and minorities filling the seats left empty by the decline in white male enrollments—indeed, some feminist scholars such as Wajcman and Cockburn insisted physics and engineering exhibited a clear bias against women—I was confident that I would find no gross hostility to the idea of women and minorities in a STEM discipline. Very simply, if such intense hostility existed, I would have few or no women and/or minority respondents, as they would not have entered graduate schools under such conditions. Therefore, the question I would be addressing was one of degree of welcome.

Were there any clues in the literature to what characteristics of women might affect the degree to which they felt welcome? Women who were self reliant, athletic—in short, somewhat masculine or tomboyish—probably would find graduate school most welcoming based on the literature. For instance the feminist scholars, Cockburn (1985) and Wajcman (1995) argued that "engineering projects a heavily masculine image hostile to women" and they blamed this hostility to women for the "shortage of women engineer role models to act as a buffering influence and mentor other women into the field." (see Fisher, 2007, p.21 and also references quoted therein). Perhaps, however, Cockburn and Wajcman exaggerated the hostility to women in engineering. Perhaps engineering was more hostile to a particular form of femininity, the shrinking violet sort of women, rather than to the presence of women per se.

Indeed, Etzkowitz et al.'s assertion (p.74), "Contrary to gender stereotypes, female graduate students are often left to be 'rugged individualists,' having to fend for themselves " indicates that women who are in fact self reliant, confident individuals could thrive in the graduate school STEM programs. It is only women who are more dependent emotionally who find the schools unwelcoming. Of course, white men, even those who want more emotional support and encouragement, find the graduate school environment congenial as Etzkowitz et al. point out. These white male students are usually drawn into the coterie of people surrounding a professor. The "caring, supportive environment" of this professor's coterie mitigates the worst effects of the weed-out system for men who survive graduate school in greater proportions than women.

If some prior research suggested that women of a self reliant stripe might thrive in the STEM programs in the graduate schools, other work hinted that women were "tracked" into career paths that were less remunerative, less likely to lead to great success in the discipline. For instance, Xie and Schauman, 1998, 2003, summarized in Fisher, 2007, p. 46-49, showed that women professors usually had more burdensome teaching loads than men during the period 1969-1993. Valian pointed out that women were more likely to be represented in the teaching oriented, rather than the more prestigious research oriented, universities and colleges. Valian further says women are "more highly represented among those who do not publish at all."[1] Since graduate schools are primarily training researchers, the fact that women faculty are more likely not to be publishing, to be faculty in the teaching oriented schools, raises the possibility that women, far more than men, are steered in the teaching direction while in graduate school.

If women were being steered towards teaching (or at least away from research) while still in graduate school I wondered if I would find that women graduate students had more burdensome undergraduate educational duties than men. And, as I looked at the various ways schools could encourage (or discourage students), I also wondered whether differences in ethnic backgrounds of the women would affect the degree to which the women would report feeling encouraged or discouraged.

I could find no studies that provided a firm theoretical basis for expecting the graduate schools in general to prefer one group of women. Different criteria might lead to differing arrays of students by faculty preference. Was social class the most relevant criterion? Then all else being equal the school might encourage upper class Hispanic men more than talented Asian women from families of modest means to succeed.

On the other hand, if the number of traits in common with the faculty were the criterion, perhaps a wealthy talented Asian woman would be preferred by the faculty to an equally able Hispanic man who came from modest means. A white woman from an Ivy League college might be preferred to a white man from a state university.

In this chapter I look more closely at women graduate students, in particular, at women of African ancestry and Hispanic heritage in contrast to the previous chapter where I focused on men, in particular African ancestry and Hispanic heritage men compared to white men. My purpose her is to provide a more nuanced view of what the graduate schools do in regard to women's graduate education than the reader would gain from alone. There, my purpose was simply to emphasize how well the schools treated white men students compared to African ancestry and Hispanic heritage men in particular but more generally everyone other than white male students. Yet, in making that point I risked fostering the impression that somehow compared to white men, everyone else was in the same boat.

It is important to understand that women are not treated all the same. Some women, especially white well educated women, are more encouraged by the graduate schools to complete their doctoral studies than others. In a preliminary way, I pointed to this in my previous monograph.

The point is critical. If graduate schools treated all women the same, then one set of policies may be required to level the playing field for women compared to men. On the other hand, a different set of policies needs to be considered if how the schools treat women depends on their ethnic background and social class. For example, if all women were treated the same by the graduate schools—e.g., the graduate schools look at women students as not seriously interested in pursuing degrees because they all want to start families—perhaps the solution is to offer day care for the children of women who might otherwise drop out to start families.

However, if the graduate schools have different views of women depending—about getting degrees—willing to place their young children in daycare while the mothers are studying—whereas the schools believe African ancestry and Hispanic heritage women are not serious about their studies—because the schools believe women of these latter two groups feel they should be home with the children once they start their families—then clearly providing daycare is not the whole answer. Other steps must be taken if the faculty and deans have differing perceptions of the way white women regard day care from the way African ancestry women see the value of daycare for their children.

What might these steps be? Perhaps, finding star faculty of Hispanic heritage and/or African ancestry is one possibility. Another might be recruiting more students from African societies below the Sahara ("Sahel" nations) who come from societies where African people are in top positions in the academic world etc.

Similarly, for instance, if white women are more likely to be willing to become part of the coterie around a professor than other women, for instance Asian women, then the schools need a different set of policies than if all women feel the same about hanging out socially with a professor and his graduate students.

If the focus of the previous monograph were women, whereas the focus here is African ancestry and Hispanic heritage students, what will I discuss now that I have made the point that white men are better treated than African ancestry and Hispanic heritage men. And I also showed in Chapter Two that Asian men are better treated than African ancestry and Hispanic heritage men.

My answer is simple. Thanks to the addition of a survey of African ancestry and Hispanic heritage graduate students I am now able to address the situation of women of Hispanic heritage and African ancestry not just in relation to white men and Asian men but in relation to other women. Furthermore, I can also present additional analyses of school milieu that should be valuable in understanding the situation of women graduate students. It seems pertinent, therefore, to revisit the situation of the female graduate students I initially explored in the previous monograph and in this chapter I will do that.

Direct Encouragement of Women

In this section, I look at indicators of direct encouragement of students. I am concerned with how well African ancestry women and Hispanic heritage women fare compared to women of white and Asian cultural backgrounds.

One caveat is necessary in reading these tables. I am comparing white and Asian women from a select group of schools in the Northeast, the South, and the West with a respondent group of African ancestry and Hispanic heritage women from sixty schools across the United States. It would be incorrect, for example, to say that University X treats its Hispanic heritage women and/or African ancestry women differently (better or worse) on the basis of these data than it does its other students, either female students or male students. What the data show, however, is that the overall national situation of the few African ancestry women and Hispanic heritage women in the sciences and engineering needs to change, both to improve the experience of the current group of students and to encourage more to come. Beyond that, it is necessary for each school to look at its own situation and see how best to improve the situation of its students.

The first issue I want to examine is feelings of general encouragement. Table 3.1 shows just 26.2 percent of the African ancestry women and 32.2 percent of the Hispanic heritage women felt generally encouraged. This compares to 47 percent of the white women and 49.6 percent of the Asian women. Even compared to other women, it is clear that African ancestry and Hispanic heritage women students did not feel very encouraged to succeed. This finding came as quite a shock. Yet even more surprising, I found that African ancestry and Hispanic heritage women were quite satisfied with their graduate school experience. In fact, the satisfaction scores of African ancestry women (4.00) and Hispanic heritage women (4.08) were actually higher than those of Asian women (3.73) and white women (3.83). The difference was statistically significant (p=.077, df=3 using the Kruskal-Wallis equality of populations rank test).

Table 3.1: Encouraged to Pursue Doctorate by Student Background for Females

Student Background	Yes	No	Additive Index	Factor Analysis
White Female	47.0 (141)	53.0 (159)	.5748 (300)	.1413 (300)
Asian Female	49.6 (56)	50.4 (57)	.5926 (113)	.1907 (113)
Black Female	26.2 (11)	73.8 (31)	.3839 (42)	-.5208 (42)
Hispanic Female	32.2 (29)	67.8 (61)	.3597 (90)	-.5819 (90)
Total	236	307	545	545
Kruskall-Wallis equality-of-populations rank test	Pearson chi-square test: Chi-square=13.048, p=.005, df=3		Chi-square with ties = 45.341, p<.000, df=3	Chi-square with ties = 46.807, p<.000, df=3

What then accounts for the satisfaction? Part of the answer may be that sample attrition was at work here. Whereas 545 women answered the question regarding encouragement, 513 or about 6 percent fewer answered the question about satisfaction.

I am not entirely convinced that this is the whole answer, however. It is plausible that the African ancestry and Hispanic heritage women have different expectations than white women and Asian women and the satisfaction scores reflect this. Suppose the African ancestry women and Hispanic heritage women expected much worse treatment than they actually got. Suppose, in particular, they did not expect any encouragement to succeed. Indeed they came fearing they would be mistreated. If their fears were not realized and their low expectations were met, they might feel quite satisfied and yet also report truthfully that they were not encouraged to succeed.

White women and Asian women, expecting more than they received in the way of encouragement to succeed, might be a less satisfied than they expected to be. But since about half of them also conceded that they were encouraged to succeed, these more encouraged to succeed might see in the treatment they received reason for satisfaction. Hence, they would lift the overall satisfaction scores for their ethnic group.

I have no way to test these hypotheses. However, I can examine whether the differences in feelings of general encouragement between white women and Asian women on one hand and African ancestry and Hispanic heritage women on the other are also evident when data on specific encouragement are considered.

Table 3.2 shows that white (45.3 percent) and Asian women (49.6 percent) believe their graduate school does bring in guest speakers to talk about job opportunities in the discipline more than African ancestry (33.3 percent) and Hispanic heritage women (23.3 percent) do. While the differences are significant, this table hardly is a rousing endorsement by any of the women graduate student groups of the practices of the graduate schools since in no case did a majority of the students assert their schools bring in guest speakers. Perhaps the graduate schools should give more priority to bringing in guest speakers—both speakers from minority groups and women speakers—to discuss opportunities in the field for successful graduates.

Table 3.2: School Brings in Speakers to Talk About Job Opportunities by Student Background for Females

Student Background	Yes	No	Total
White Female	45.3 (136)	54.7 (164)	300
Asian Female	49.6 (56)	50.4 (57)	113
Black Female	33.3 (14)	66.7 (28)	42
Hispanic Female	23.3 (21)	76.7 (69)	90
Total	227	318	545

Pearson chi-square test: Chi-square=18.202, p<.000, df=3

One area where it is clear that schools treat some women better than others is financial aid. Eighty-four percent of the white women in the respondent sample compared to 47.6 percent of the African ancestry women and 57.8 percent of the Hispanic heritage women reported that the school provided financial aid (see Table 3.3 below).

Table 3.3: School Provides Financial Aid by Student Background for Females

Student Background	Yes	No	Total
White Female	84.0 (252)	16.0 (48)	300
Asian Female	78.8 (89)	21.2 (24)	113
Black Female	47.6 (20)	52.4 (22)	42
Hispanic Female	57.8 (52)	42.2 (38)	90
Total	413	132	545

Pearson chi-square test: Chi-square=45.630, p<.000, df=3

Table 3.4 shows that white (78.3 percent) and Asian women students (77.9 percent) far more than African ancestry (40.5 percent) and Hispanic heritage women (40.0 percent) assert their school provides opportunities to work on research problems, good practice for their futures as research workers in their respective disciplines.

Table 3.4: School Provides Opportunities to Work on Research Problems for Females

Student Background	Yes	No	Total
White Female	78.3 (235)	21.7 (65)	300
Asian Female	77.9 (88)	22.1 (25)	113
Black Female	40.5 (17)	59.5 (25)	42
Hispanic Female	40.0 (36)	60.0 (54)	90
Total	376	169	545

Pearson chi-square test: Chi-square=67.730, p<.000, df=3

Regardless of ethnic background, the majority of women graduate students complain about the heavy undergraduate teaching responsibilities imposed by teaching assistantships. However, as Table 3.5 indicates, African ancestry (71.4 percent) and Hispanic heritage women (75.6 percent) complain far more than white women (58.3 percent) or Asian women (54.9 percent) in this regard.

Table 3.5: Assistantships Do Not Require Heavy Responsibilities for Teaching Undergraduates by Student Background for Females

Student Background	Agree	Disagree	Total
White Female	41.7 (125)	58.3 (175)	300
Asian Female	45.1 (51)	54.9 (62)	113
Black Female	28.6 (12)	71.4 (30)	42
Hispanic Female	24.4 (22)	75.6 (68)	90
Total	210	335	545

Pearson chi-square test: Chi-square=12.624, p=.006, df=3

Among women only a minority assert that their school provides them with access to the latest equipment for their research. Yet, Hispanic heritage women are less likely (28.9 percent) than any other group—including African ancestry women (42.9 percent) to claim the school provides them with access to the latest equipment for research (see Table 3.6 below).

Table 3.6: School Provides Access to Latest Equipment by Student Background for Females

Student Background	Yes	No	Total
White Female	46.3 (139)	53.7 (161)	300
Asian Female	45.1	54.9	113

Student Background	Yes	No	Total
	(51)	(62)	
Black Female	42.9 (18)	57.1 (24)	42
Hispanic Female	28.9 (26)	71.1 (64)	90
Total	234	311	545

Pearson chi-square test: Chi-square=8.884, p=.031, df=3

Before I move on to the next topic, I want to briefly discuss how African ancestry women and men compare to each other and how Hispanic heritage men and women compare to each other.

In regard to general encouragement, I found no statistically significant differences between African ancestry men (31.3 percent) and women (26.2 percent) with only a small minority feeling generally encouraged to complete their degrees successfully. The same held true for Hispanic heritage men (33.6 percent) and Hispanic heritage women (32.2 percent).

The following chart shows that when you examine gender differences within each of the two ethnic groups, they are not statistically significant and in only a couple of instances can this be partly a result of small sample sizes.

Table 3.7: Within Ethnic Group Percentages on Specific Encouragement Variables

Encouragement Variables	Black Male	Black Female	Hispanic Male	Hispanic Female
School brings in speakers	21.9	33.3	26.7	23.3
School provides financial aid	53.1	47.6	60.3	57.8
Opportunity for Research	50.0	40.5	48.3	40.0
Assistantships impose heavy undergrad teaching load	71.9	71.4	72.4	75.6
Encouraged to start/continue research	37.5	40.5	31.9	38.9
School hires good profs	40.6	47.6	40.5	42.2
Provides access to latest equipment	34.4	49.2	37.9	28.9

No statistically significant differences were detected within ethnic groups at the .05 criterion. The chi-square test was employed on these data.

SCHOOL ATMOSPHERE

The main question this study addresses is what do universities do to encourage their graduate students to complete their doctoral studies successfully and go on to rewarding careers in their chosen field? Thus far, I have looked at the steps graduate schools take that directly encourage graduate students—providing financial aid, bringing in speakers to talk about opportunities in the field, providing opportunities to work on real problems in the discipline that need solving, and so forth.

Schools do other things that indirectly encourage or discourage students from completing their doctorates as well. The kind of milieu a school fosters can be important. In CATSG I found, for example that "women in a friendly atmosphere . . . felt more encouraged to pursue their doctorates" (see CATG, p. 91). I also pointed to unsettling evidence of sexism among graduate students, primarily among the Asian students who are a major component of graduate school students especially in the hard sciences.

In the current study I revisit the milieu issue because my discussion of this issue was incomplete in CATSG. Before I introduce the material that has not previously been presented, however, I want to briefly recapitulate some key findings from CATSG.

Sociologists who have studied graduate school experiences of students have argued that the contacts the students make may be critical to their subsequent career success. Women, for example, have fewer research collaborators than men according to Bozeman and Lee (2003) and I maintain this problem begins in graduate school (CATSG, p. 137-138).

One of the tantalizing findings is that more often than the other male students, Hispanic heritage and African ancestry male students assert, "connecting with other graduate students is important for success in graduate school" and "contacts I made in graduate school will be helpful in my future career" (see Table 3.8 below). Why this is so is not readily apparent. Possibly the African ancestry and Hispanic heritage students think that the reason they are not doing as well, not as likely to be granted doctoral candidacy, not as likely to finish their degree studies successfully as their white and Asian fellow students is related to their inability to forge useful contacts. And in their perceptions they may be somewhat correct although this is not determinable from the data I possess.

Table 3.8: Connecting With Other Graduate Students is Important for Success in Graduate School by Student Background

Student Background	Connecting is Important	Total
White Male	3.30	293
Asian Male	3.28	154
Black Male	3.39	31
Hispanic Male	3.60	114
Female	3.26	534
Total		1126
Kruskall-Wallis equality-of-populations rank test	Chi-square with ties = 14.943, p=.005, df=4	

Among women students, connecting with other students or belief that contacts made in graduate school will be helpful in their future careers is not regarded by Hispanic graduate students as any more significant than among other women students. The difference between Hispanic men and Hispanic women on the importance of connecting with other graduate students is statistically significant as indicated in Table 3.9 below.

Table 3.9: Connecting With Other Graduate Students is Important for Success in Graduate School by Student Background for Hispanic Students

Student Background	Connecting is Important	Total
Hispanic Male	3.60	114
Hispanic Female	3.34	88
Total		202

Two-sample Wilcoxon rank-sum test, Z=-2.058, p=.040

Hispanic men students also were more likely than any other students to claim the graduate school atmosphere was "competitive" (see Table 3.10). Is there a relationship between the two feelings? It is certainly plausible-that feeling the school atmosphere is competitive and hence very demanding intellectually might be behind a desire for connecting with other graduate students to be successful. Students under these circumstances would want to study together, work together on solving problems that might be baffling and overwhelming for any one of them individually.

Table 3.10: Competitive Atmosphere by Student Background

Student Background	Strongly Agree/Agree	All Others	Total
White Male	47.5 (142)	52.5 (157)	299
White Male: CO	50.9 (27)	49.1 (26)	53
White Male: NYC	47.4 (54)	52.6 (60)	114
White Male: NC	45.8 (60)	54.2 (71)	131
Asian Male	63.9 (101)	36.1 (57)	158
Black Male	53.1 (17)	46.9 (15)	32
Black Male: Native	52.6 (10)	47.4 (9)	19
Black Male: Foreign	53.8 (7)	46.2 (6)	13
Hispanic Male	69.0 (80)	31.0 (36)	116
Female	58.7 (320)	41.3 (225)	545
Total	660	490	1150

Pearson chi-square test: Chi-square=21.724, p<.000, df=4

To investigate this possibility I ran several cross tabulations both to determine if a relationship indeed existed between competitive atmosphere and desire to connect with other graduate students to ensure success on one hand and to determine if it was stronger or only present in some components of the study sample. However, the results were not statistically significant.

Yet milieu of the university is important as I discovered when I looked at how results varied for the relationship between "class mates think men make better professors, scientists, and team members" or "men make better professors, scientists, and team members" on one hand and professors' respect for ethnic and gender differences (as assessed by the students) on the other among women respondents.

This analysis of sexism among students is extensive and I deem the subject sufficiently important that it deserves a separate chapter by itself. Therefore, I will defer to the next chapter this discussion and summarize the present chapter's finings.

SUMMARY AND CONCLUSIONS

The main theme I have emphasized in this book has been that white male students have the best graduate school experience of any student group. White male students have the best financial support; are more likely to find their professors knowledgeable about the subject matter and encouraging of the students' research work; are more likely to claim they have access to the latest equipment for their research.

This is not to say that white male students regard graduate school as a nirvana. They grumble about the burdensome undergraduate teaching responsibilities of assistantships. They face long years of research work to get their doctorates before they embark on their careers. And in these uncertain economic times even when they receive their doctorates they do not necessarily land good paying jobs. Much depends on the discipline in which they receive their doctorates with fields such as chemistry and mathematics and engineering offering the best job prospects. Social science graduates face bleaker prospects and in some specialties of biology job prospects are also not attractive.

If it is true that relatively speaking white male students have a rewarding experience, it is less true of other students. However, as the data in this chapter show, and as was suggested in some findings presented in my previous work, women's experience is not uniformly dismal. Generally white women enjoy a relatively superior experience to other women and once they graduate with their doctorates they may find rewarding careers in specialties in demand. Of course those in specialties in less demand will face prospects as dismal as male students in those specialties.

Asian women, Hispanic heritage women, and African ancestry women, however, are not doing as well as white women. The Hispanic heritage and African ancestry women have much worse experiences in graduate schools than white men, and even compared to white women have worse experiences. For instance, as was shown above, the African ancestry women and Hispanic heritage women are more likely to have assistantships with burdensome undergraduate educational responsibilities. They are less likely to have access to the latest equipment for their research. And any women who are not part of the coterie around a professor—and this may well include the Asian women and Hispanic women far more than white women—will not forge the informal contacts, nor hear about the latest research ideas that are subtle but essential advantages they will need to have good careers in science and engineering.

We must reform assistantships drastically to emphasize them as tools for educating students rather than as tools for exploiting their labor. Assistantships usually go to the less promising students at a university but less promising is not equivalent to unpromising. A student who eagerly accepts an assistantship at a prestigious graduate school such as the University of California at Berkeley may be less promising than the student who receives a fellowship from that university. However, that recipient of an assistantship may well be as promising as the fellowship recipient at a less prestigious school. Driving that student out of the program because the student was only valued as cheap labor in undergraduate courses—frequently the case in the recent past with students enrolled at prestigious schools such as Columbia University as well as less prestigious schools—is an appalling waste of scientific and engineering talent and simply indefensible in these times. And the fact this weed out process probably happens more often with women and minority students than with white male students could well eventuate in class action lawsuits that might only compound the financial woes of the cash strapped universities.

In the next chapter, I take up an issue touched on in CATSG that affects students regardless of their gender and ethnic preference although it probably affects women and minority students more than white male students. The issue is the influence of a climate that is perceived by the students as sexist or racist on the sexism and racist sentiments of the students. In my opinion, when students successfully complete their studies while holding these sexist or racist sentiments, they will perpetuate the resistance of graduate schools to increasing the numbers of women and minority faculty and/or improving the educational experience of women and minority students. Moreover, when the sexist views are held by women, I believe they serve to drive these women out of science and engineering. This is especially true when they work in tandem with pressures from spouses to set aside the woman's own science career in order to assume family burdens and/or advance the husband's career without regard to the wife's own career.

Endnotes to Chapter Three

1. Valian, reviewing findings of prior research, pointed out "If women are less successful than men at getting tenure-track jobs at elite institutions—as they are—more women than men will be unproductive." She concluded this would occur because "the prestige and resources—such as equipment, library holdings, support staff, and internal research grants—of a scholar's academic institution influence productivity more than productivity influences the chances of getting a job at prestigious institution (p.263)." Valian also makes the interesting observation "the culture at prestigious universities does not just help scholars be more productive than they would be at less elite institutions, it forces them to be." And she further suggested, "if the prestige of the institution plays a role in determining productivity, then men and women in similar ranks at elite institutions should be roughly equal in productivity (p.263)." Valian was referring, of course, to faculty but her observations could possibly apply equally to students at these prestigious institutions.

Chapter Four
Are (White) Men Better Professors and Scientists?

The Persistence of Sexism Among Graduate Students in Science and Engineering: Its Causes, and What We Can Do About It

American science is mostly white male science. In 2005, Pinker pointed out that women comprise no more than ten percent of scientists and engineers in academia and Megaw (1991) showed that in physics they comprise perhaps as little as four percent of all academic faculty.

That American science is a largely male world is mainly because the universities in which research is conducted are largely white male institutions. Universities in the United States began in the nineteenth century in places such as Harvard and Columbia—both male colleges—and took as their models German (and, to a lesser extent, British) universities that excluded women and African Americans from the student body as well as the faculty (see Ehrenreich and English, 1973; Epstein, 1970; Rossiter, 1982; Fisher, 2007). In the words of Cynthia Fuchs Epstein, 1970, locating science in the university setting "while permitting concentration of tasks and resources also restricts the pool of scientific personnel to the like-minded and similarly experienced (p. 241).

Therefore, until recently, although women have contributed to science from earliest times (e.g., the Marquise de Chatelet in eighteenth century France (see Haier, 2007, p.117; also Osen, 1974), Madame Curie in the nineteenth and early twentieth century, Barbara McClintock and Linda Buck late twentieth century and/or beginning of the twenty-first century), they have been discouraged from pursuing science as a vocation. However, the demographic profile of American science is changing, as is the demographic profile of American professions generally. Women now are university presidents (e.g., Shirley Jackson of Rensselaer Polytechnic Institute, Shirley Tilghman of Princeton University (until recently), and Drew Gilpın Faust of Harvard); deans of graduate faculty at prestigious engineering schools such as Massachusetts Institute of Technology; sit on the Supreme Court of the United States; and have penetrated virtually every occupation practiced in the United States.

The notable increase in the number of women in professional occupations over the past few decades and the accompanying change in the demographic profile of professions in America is not a temporary aberration. More women than men receive bachelor's degrees and women now earn 75 percent of the veterinary medicine degrees and about half of the medical degrees. As Diane Halpern says, "It is clear that women are succeeding in some sciences and thus must have the cognitive ability to learn and succeed in science" (Halpern, 2007).

Attitudes that women "do not belong in science" also are changing. When, in 2005, Lawrence Summers, then President of Harvard University, suggested women were not more numerous on the faculty at prestigious research universities such as Harvard University because they did not have the correct traits to do work at the frontiers of science, his comments ignited an uproar and he recanted his views.

Not only are American attitudes about women's ability to do science more positive, my survey results also show overall sexist attitudes among S & E graduate students are uncommon, with no more than five to ten percent of the overall sample having sexist attitudes. Sexist attitudes are especially uncommon among white students—the large majority of graduate students—and African ancestry students. This is demonstrated in the fact that the vast majority of the graduate students in this study answered that they did not agree that men make better professors, or better scientists and engineers, or better team members than women.

It would be a mistake, however, to see in these encouraging results the imminent demise of sexism. (And even if it were, a combination of factors such as no required age of retirement will keep women out of the top tier of academic positions for decades to come (see Kaminski, Deborah and Cheryl Geisler, 2012). While the evidence is strong that sexism is a waning problem in the United States, the fact remains that it has not been eradicated and may not be eradicated any time soon if indeed it ever will.

Defining Sexism

Up to now I have been talking about sexism as if everyone knows what it is. However, if we are to measure its extent, and its correlates (if not its causes) we need to have a clear, operational definition of the term. Sexism, broadly speaking, is a belief that one gender is somehow superior to the other and the inferiority of the other gender accounts for its lesser attainments and status in society.

It is important to understand, however, that sexism may manifest itself differently in different contexts. In the executive suite, it may manifest itself in women being expected to be the note takers at meetings, fetch the coffee, and so on. In science, however, it will more likely be demonstrated in the relative lack of collaborators on research projects that women researchers have, or in the belief that women lack the necessary mental attributes such as "spatial reasoning" to do good physics. And these beliefs will be highly correlated with such consequences as women's slow promotion to the ranks of senior professors, and the like.

However, it would be wrong to rush to judgment when we see a situation of women at a particular university department not attaining senior rank as quickly as men do. It is certainly possible that the men are much better than the women in that department as evidenced by more and more important publications and so on. Buckley et al. (2000) suggest in their study that the differences between men and women physicians in publication productivity are not statistically significant but I question their relying solely on the chi-square test of significance to reach their conclusion and their ignoring other statistical measures such as effect size differences. This, in fact, is a key point that Cole and Singer made in their "theory of limited differences." According to Cole and Singer, the rewards differential between men and women in science once they embark on their careers reflects real differences in accomplishments. Cole and Singer see a relatively uniform pressure on girls beginning in their childhood and extending through graduate school years either to avoid science altogether or at least to give higher priority to family concerns than to their careers. In contrast, I have emphasized the importance of graduate school training differences as key to understanding the differential success of men and women once they embark on their careers in science and engineering although I acknowledge prior socialization can be important in some instances as well

If absence or near absence of one gender from a particular occupation is not by itself proof of sexism, it is nevertheless highly correlated with sexism. And if the rationale for the absence or near absence of women from a discipline or group of related disciplines, for example, is found wanting intellectually, then we must conclude that sexism plays a role in perpetuating their absence.

In my opinion, sexism is a major factor in the absence or near absence of women from science and engineering and especially from senior positions in the academic world. But as I said above sexism in academia does not necessarily manifest itself in the same way as sexism in the trade of bricklaying or among the steelworkers who work on high-rise buildings, standing atop steel beams hundreds of feet up in the air. For example, you might find bricklayers passionately arguing that women cannot work as fast in such a field as men or cannot lift bricks as well because their hands are smaller. Academics, however, might insist that in their discipline spatial reasoning ability is essential and few if any women excel in this kind of mental ability, accounting for their absence from the discipline (see Steve Hsu, 2012).

Let me be quite clear. If in fact everyone in the discipline needed to use spatial reasoning ability for all the work of the discipline then certainly we might expect a preponderance of people with that skill. However, we find on close inspection that only certain research tasks tap spatial reasoning ability.

On the other hand, all academicians must do committee work, much writing, and collaboration with people with diverse skills and backgrounds to arrange financing for projects, recruit personnel, and so on. These other tasks that faculty members engage in require a different group of skills than spatial reasoning ability: social competence, verbal skills, etc., where women do quite well.

My point is that an effective department of mechanical engineering where, I concede, research may require spatial reasoning skills of a high degree, also needs a variety of other skills. Furthermore, only some people in the department require a high degree of skill in spatial reasoning whereas many may only require some spatial reasoning ability to teach the subject.

In sum, it seems difficult to believe that women in the United States could not perform well in academic engineering and physics departments when they are able to do this in countries such as France, Italy, Russia, and Hungary (to name just a few). What is it about physics and engineering in the United States compared to these other nations that in the American academic environment women are virtually absent from fields where spatial reasoning ability is a major asset?

While I think sexism is a major factor in the persistence of a lopsided ratio of men to women in S and E departments, I do not think sexism is a major problem in American science and engineering on the order of inadequate numbers of well trained high school science teachers who not only know their science but can teach science in a way that makes it interesting for students.

If sexism were a disease, I would consider it more akin to a health problem such as Legionnaire's disease than to influenza. Simple effective remedies exist to control Legionnaires disease it but it can flare up when schools become blasé about it. And just as Legionnaire's disease is a potentially lethal illness that can sicken numerous people if not diagnosed and treated properly and promptly, sexism once it flares up on a campus can blight the lives and careers of numerous people much as racism or anti-gay attitudes can.

Why is it that if sexism is discredited it will not disappear? Unfortunately, in my opinion, sexism is nourished by three interacting factors: (1) an influx of foreign students from cultures where women are expected to be deferential to men; (2) the continued respectability of research purporting to show that women's psychological traits somehow are causing the gender gap in science, and (3) the ability of "star" faculty members—nearly always men—to recruit and reward students they want to help in their careers.

I want to briefly comment on items 2 and 3 before amplifying my views about the naturist tradition of research on the gender gap in science. I will begin with the influence of star professors on student sexism.

I maintain that sexism is nourished by the simple fact that the senior faculty of a department is overwhelmingly masculine. Perhaps more important is the fact that the leading professors, the ones who control resources and have contacts in other prestigious departments, are almost always men. Thus, even if a token woman is present, the fact that leading men professors control more resources, can offer their students richer experiences than women professors (if the men professors choose to) subtly reinforces any sexist attitudes the male and female students have. The reason is simply that to the impressionable young graduate students, what is clear is that men command the resources for research. Men professors have the contacts in other schools to help their best students land the most desirable junior academic appointments such as prestigious postdoctoral fellowships and assistant professorships at the best universities. Men professors are in the discussion groups and networks where the most important research questions are discussed and ideas exchanged. Men professors snare the most promising students as research assistants.

These feelings that men professors are superior, men scientists are superior, and men team members are superior help perpetuate the lopsided ratio of men to women in top scientific positions. This is less likely to happen, in my opinion, when the school also has a number of prestigious senior women professors in the same departments as top male professors.

Another factor in the durability of sexist sentiments is a large influx of foreign students from cultures where women are expected to be deferential to men. Students from these cultures can create a climate in a department that is unfriendly for the women if the school is lax in insisting on a gender-neutral policy where persons of ability regardless of race, religion, or gender are welcome. Up to thirty percent of these students from Asia may exhibit some sexist attitudes according to my data (See CATSG).

The overwhelmingly male milieu of many graduate S & E departments, combined with a large influx of foreign graduate students, has potential to create a sexist atmosphere. To put it bluntly, the little emperors from China—and their counterparts from India, Iran, Arab lands and so on—arrive on our campuses with attitudes that are deeply ingrained in their societies. This is perhaps understandable. What is worrisome is the possibility that university administrators, desperate for income to meet their budgetary needs, may overlook this reality in their eagerness to attract these high paying graduate students from China or India or the Middle East. School administrators need to be reminded sometimes that the students from abroad are here to learn our best practices and ideas and not to teach us their worst practices and attitudes. If the university administrators are vigorous in getting out the message that all who seek an education are welcome, then a flare-up of sexism can be avoided in my opinion. Otherwise, a chilly climate for women graduate students can develop quickly and with unfortunate consequences.

I have focused up to now on male attitudes of sexism and how they might arise in societies that expect women to be deferential to men and also to be nurtured in the highly masculine world of the American university, especially the science and engineering departments. It is important, from both a policy and theoretical standpoint, to point out perhaps women in the S and E disciplines also share the feelings that men make better professors, or make better scientists and engineers, or team members on research projects.

Women who hold these beliefs in my opinion are vulnerable to pressures to abandon their studies altogether, to settle for part time work, to give up trying to reach the top echelons of the discipline. Without saying explicitly that it was internalized sexist attitudes that men were better professors, or better scientists than women (especially themselves). Cole and Singer emphasized that the women's reactions to setbacks, rather than their setbacks per se, were much more important in understanding why women scientists and engineers failed to be as productive as men in their careers after they received their doctoral degrees. These women, if I am correct, magnify the importance of setbacks, inevitable in any arduous course of study, to reach the erroneous conclusion that they will never succeed no matter what, rather than resolving to work harder, to give more attention to their studies and spend less time on distractions that interfere with their mastery of their field.

In many, if not most cultures, women may find giving up their quest to become scientists, certainly their quest to become top professionals, is easy because they can gain social approval for focusing on becoming mothers, emphasizing their domestic responsibilities. Spouses will not disapprove, indeed may be pleased that the women do this. Parents, and parents of the spouse, will likewise support such a decision since in many cultures it is important to produce male children and the spouses and the parents of the couple will desire grandsons. Men in contrast would not gain social approval from failing to succeed as professionals. Men expect themselves and feel others expect them to be ambitious; to make a good living that supports their family in comfort. Lacking an easy alternative to competing for success in the world of work, an alternative that brings societal approval as it does for women who elect to be mothers, men are more likely in the face of a setback to persevere and succeed.

Evidence that women often embrace motherhood and family responsibilities rather than struggle to succeed in science was adduced by Williams and Ceci (2012) in the literature review they did of psychological writings on the gender gap. I had earlier (in a rejoinder to Williams and Ceci, submitted to The American Scientist) suggested that women, rather than abandoning their careers to start families, might be leaving because of the pressure to care for ailing or enfeebled parents (and parents of their spouses). I still believe this may be part of the explanation of what is happening. However, another source of pressure, it now appears, is the sexism that women have internalized. This would help explain women leaving science while in college, or while in graduate school, when the parents are probably too young to require the daughters' assistance in caring for them.

Internalized sexism affects "self efficacy" in my opinion. Leslie et al. (1998) point out that "as specific obstacles arise, one's self-efficacy will be instrumental in determining whether one decides to cope with adversity, to what degree, and how persistently" (p.253). Bandura, to whom we owe the theory that Cole and Singer rely on for their model of "limited differences,"—the model that is adopted in general for this study—has written that the "strength of conviction" in your self-efficacy will affect not only your willingness to see a task through to completion but even whether you will make an effort to do the work. As Bandura puts it,

> Weak expectations are easily extinguishable by disconfirming experiences, whereas individuals who possess strong expectations of mastery will persevere in their coping efforts despite disconfirming experiences. (Bandura, 1977, quoted in Leslie et al., p. 253).

The point of view Bandura advocates, and that is shared by Leslie et al., suggests that Bandura's views hark back to an earlier formulation by Leon Festinger that he referred to as "cognitive dissonance theory." That theory states that once a decision has been made the evidence in favor of that decision becomes salient while evidence against it diminishes in importance and recall. In other words, one plays down disconfirming evidence. (On cognitive dissonance theory, see e.g., Festinger, Leon, 1962).

The "internalized sexism" I hypothesize as a mechanism for pushing women out of science early in their training and careers is essentially the point made by Cole and Singer (1992), who in explaining why men publish more scientific papers than women, emphasized the various slights and setbacks women endured and how women would react to them. (Not surprisingly, Cole and Singer embraced Bandura's psychological theory by name in proposing their own model of why women leave science or produce less in their careers than men).

Cole and Singer's ambitious mathematical model of how the gender gap in science would arise and worsen over time has never been fully tested. I made a start on this in my earlier monograph CATSG by showing that differences in perceptions of the graduate school experience among a sample of graduate students suggested that graduate schooling experiences did have a probable effect, as the Cole and Singer expected, on subsequent careers of white men compared to women and Asians who received doctoral degrees.

However, seen from the vantage point of twenty years of work in the field subsequent to their theoretical contribution, it is clear Cole and Singer were very astute, even prescient, in predicting that (1) women might drop out of graduate school and/or choose to work part time after receiving their terminal degrees because of the accumulated weight of prior socialization experiences, as opposed to only those they had in graduate school—which my own research has emphasized. If that prior socialization made them vulnerable to the pressures to depart science emanating from their present environment even if their experiences in graduate school were generally positive, they would flee, perhaps, although that reaction would not be expected of them based on understanding their graduate school experiences alone.

Cole and Singer did not focus on data about students from abroad in their seminal article—not surprising since data on such students was thin to nonexistent when they did their paper in the early 1990s—but this does not detract from the value of their contribution to social science theory about the gender gap in science and engineering.

The landscape has changed since Cole and Singer did their seminal work. Foreigners, especially Asian graduate students, are a prominent part of the graduate student body and the diverse cultures now present among the student body allows one to see how background, as opposed to graduate student experiences themselves, may influence women's exiting the pipeline leading to a doctoral degree in science and engineering.

NATURIST EXPLANATIONS OF WOMEN'S LIMITED REPRESENTATION IN TOP ECHELONS OF SCIENCE AND ENGINEERING

I have discussed two sources of sexism in the university: (1) the greater resources male faculty have for rewarding their students than female faculty have and (2) the influx of significant numbers of graduate students from societies where women are expected to be deferential to men.

The third factor in the persistence of sexism is the intellectual respectability of research that seeks explanations of the gender gap in science in the psychological traits, the inherited characteristics, of one gender. For example, I see sexism as receiving intellectual support from studies that demonstrate that men are somehow more suited to do science because they are more likely to have spatial rotation skills, or predominate among those who are at the top of the mathematics aptitude distribution, or are more likely to be risk takers.

I have critiqued the naturist position at length in an earlier work (Fisher, 2007) and in the first chapter of this monograph and see no reason to repeat myself now. However, I want to make a few salient points here.

First, I distinguish between the search for naturist explanations of the social phenomenon of the gender gap in science on one hand and psychological research on traits that may explain superior performance in specific specialties or in particular tasks. The former I regard as probably fruitless intellectually, as nourishing sexism, and therefore from a policy standpoint as dangerous. On the other hand, I agree with Halpern who has acknowledged that in work such as theoretical physics where spatial rotation is likely to play a significant part in one's success, it is not surprising that the most successful practitioners are people with strong spatial rotation skills (and who happen also to be mostly if not entirely male). However, until spatial rotation is shown to be strongly coupled with male sex—perhaps a neighboring gene on the same chromosome as sex—I cannot see how intellectually it is possible to defend the idea that women have no place in a discipline where spatial rotation is critical to intellectual success. And even if we come to see that in nuclear physics and certain esoteric branches of mathematics spatial rotation is essential, the fact remains that no one has shown that women cannot be department chairpersons in academic departments and no one has shown that women cannot be lead investigators in numerous hard science subjects. Lead investigators need to be effective proposal writers, people managers, and also have good research skills. Women scientists and engineers are likely to possess the skills, knowledge, and abilities required of lead investigators.

It is one of the supreme ironies that naturist studies are sometimes done by people who are genuinely committed to advancing gender equality in science. Such is the case with the paper by Williams and Ceci (2012), two very competent and productive psychologists whose views are likely to command respect. Furthermore, they published their paper in a highly respected journal, The American Scientist (2012).

The good news is that Williams and Ceci offer imaginative recommendations for steps universities could take to help women faculty members juggling the demands of marriage and family with demanding careers as research scientists and college teachers. Unfortunately these investigators made some critical errors in interpreting the results of their literature review of articles on the gender gap in science.

Illustrative of their erroneous thinking is an anecdote Williams and Ceci offer. They remark that they know a woman ("Jennifer") who "made the choice to have a family and teach mathematics part-time at a local community college." This woman, according to Williams and Ceci,

> could not imagine waiting to have children until after tenure at age 40, nor could she imagine how she would juggle caring for a young family with the omnipresent demands of an assistant professorship. The harried lives of the two tenured mothers in her department convinced her that such a path was not for her" (p. 138).

I should first concede that I do not know the woman and perhaps Williams and Ceci know her extremely well. Therefore, it is possible that the woman subject is telling the whole truth when she says she wanted to leave a prestigious university tenure track position because the harried lives of her women colleagues with children just did not suit her.

Yet while I concede the possibility that Williams and Ceci are correct in this instance, I think it implausible that most of the time we really know the full story behind the life altering decision that this woman mathematician made.

What could cause her to have given up a promising career knowing she will be leaving friends at work whom she certainly no longer see as often—if at all? More generally, she is also isolating herself from the networks in which cutting edge ideas in her discipline are discussed and dissected (see Cole, 1981). Furthermore, she knows that her income will be reduced since a part time faculty member will not earn what a full time faculty member makes (and a prestigious university likely has a higher pay scale than a community college).

And she may even be aware that her prestigious university employer offers generous maternity leave, day care, and other fringe benefits that make having a child less of a problem than it might have been if she worked for an employer who offered no such assistance. In short, she appears to be paying a steep price to voluntarily separate herself from the world of work that may forever ruin her chances to have a fine career. As I said at the outset, she chose to leave her academic life when other women she knew put up with their harried existence, hoping that eventually their sacrifice would pay off.

Williams and Ceci look at the steep price this talented woman willingly paid to be a mother and see in it evidence for a psychological drive in women to have children. I think Williams and Ceci were hasty in arguing for an exotic biological cause; probing their acquaintance's motives might well have revealed a mundane sociological explanation for her life altering decision.

In our search for a plausible explanation of Jennifer's behavior, I maintain that we might begin by looking at Jennifer's perceptions of her resources and self-efficacy as a scientist. For example, maybe Jennifer graduated from her fine university without having acquired a network of colleagues with whom to discuss the latest ideas for research. Numerous social scientists—whose works were not included in Williams and Ceci's review—have pointed to the importance of such networks including Mary Frank Fox, Jonathan Cole, Bozeman and Corley, and myself.

Alternatively, maybe, Jennifer did not see herself doing topnotch publications and overcoming the bias against women in academia. These are significant hurdles to achieving high status in her chosen profession but they are challenges that other talented women have confronted successfully.

Suppose we are able to rule out lack of confidence in her abilities as a scientist and/or membership in productive networks. Another group of motives that need to be investigated relate to her family's support for her career. In this regard, maybe, Jennifer has a husband who is very unsympathetic to her desire for a career and she wants to preserve the marriage more than she wants a career. Related to this is the possibility of her having been reared in a culture which sees the woman's place as in the home caring for the children in order to give the children the best possible upbringing. In many societies, day care is looked down on as inferior, and women who choose day care for their children are regarded as unfit mothers and poor wives. Overcoming such traditional attitudes is difficult and we should not underestimate the guilt women from these cultural backgrounds feel about not being caregivers. (It is interesting that far fewer men seem to suffer such pangs of guilt although the absent father is, I believe, a major problem in America where many boys do not have a father around who is a suitable role model of what a man is).

Another possibility behind the woman's leaving her employ is that she has other people she needs to or wants to care for and lacks the time to do this. Perhaps she has infirm parents or perhaps she has a sibling who is brain damaged and needs her help.

We have just one final point to clarify. Why did Jennifer not simply say she has to care for her infirm parents, rather than insist she wants to be a more involved in the rearing of her children than some of her colleagues? This may have to do with her desire for self-esteem and preserving her social standing in American society.

In the American culture, it is socially acceptable for a woman to be a homemaker her entire life. No one thinks less of Ms. Anne Romney, the wife of the likely Republican candidate for President of the United States, because she stayed home and reared five children as a full time mother. (Indeed when a Democratic Party operative chided Mrs. Romney, the resulting uproar led to a quick apology from President Obama.)

Yet if her husband Mitt Romney were to have made the decision to stay home and rear five children he likely would not have been a candidate for President of the United States, although aside from requiring the person seeking the office to be an American citizen by birth and at least thirty-five years of age, the most powerful office in the world has no specific criteria for selection. In American culture, the candidate for President is expected to be a man or woman who has held "real" jobs and being a father or mother simply does not count.

That "socially acceptable" excuses are not equivalent to real motives would have been clear to Williams and Ceci if they were familiar with survey research—more commonly done in sociology—and anthropological research. Unfortunately, Williams and Ceci's narrow focus on psychological literature about American (and possibly English) subjects led them to overlook the possibility that social arrangements rather than biology can explain Jennifer's behavior. (To be fair, Williams probably is aware of the anthropological literature since she has been trained in anthropology as well as psychology). Much of the psychological literature, regrettably, is a literature of white European societies such as the United States, Canada, and England. It is not a literature necessarily of studies done in India, sub-Saharan Africa and so on. This is largely because most published studies by psychologists have generally been done in advanced European societies. Replications conducted in non-western countries either do not exist or are not easily accessed by American and European psychologists.

Finally, many scientists become wedded to a particular way of establishing facts of scientific value. For example, some sociologists of the "ethno methodological" school of thought belittle survey research and champion participant observation studies. Some psychologists may be equally skeptical of survey research but champion true experimental studies. In any case, oftentimes ignorance of research done in different specialties, or by different means than the reviewer favors, can lead the scientist to misinterpret evidence or otherwise reach erroneous conclusions. This seems to have been what happened in the review Williams and Ceci published. They simply overlooked the possibility that social arrangements led to the results they were seeing after they unwittingly decided to ignore any literature that might have suggested limitations on the external validity of the impressive group of psychologically oriented studies included in their literature review.

My point is that we can suggest numerous possible reasons for Jennifer's behavior that should be eliminated only after careful inquiry. The concept of "biological drive" should be saved for those rather uncommon situations where all possibility of social causes has been ruled out. To do otherwise risks debasing this important biological concept. Not only is that bad for science it may spawn bad policy to boot. In regard to the latter, one danger is that when we see women's behavior as biologically driven we have a tendency to think that women can never be as good as men. It reminds me of Pinker's observations that women are more interested in having a life than men, or that women are less inclined to be risk takers than men so why bother trying to recruit women for top professorships in science and engineering which (Pinker believes) should go to people who care only about their work and nothing else. Pinker sees a biological basis behind these behavioral and attitudinal differences between men and women in the United States. In contrast, whereas while I agree that more women than men in the United States seek balance I maintain his point is a valid comment on current gender-based American values rather than of some fundamental and immutable biological difference.

I digress and it is important to return to the main business of this chapter. I therefore will next discuss the organization of the chapter and its findings.

CHAPTER ORGANIZATION

In this chapter I have two aims. One is to examine the extent of sexism among the S and E graduate student population. The second is to look at sexism in the female student population.

The first aim is primarily addressing the policy issue, is the university science and engineering community ready for an influx of women in faculty positions within the sciences and engineering departments? The data suggest that indeed they may be.

The second aim is to assess if women themselves are ready for this change in the university faculty. The answer again is affirmative. However, as I will show, this readiness does vary somewhat by ethnicity. Furthermore, the readiness of men and women S and E students for gender diversity is somewhat dependent on the students' perception of faculty respect for gender differences.

One final comment about the content of this chapter is germane. I spend more time on women's internalized sexism than on the distribution of sexism across the entire student body of S and E students because I want to assist university administrators and policy makers in government, academy, and industry in acquiring a deeper understanding of the policies and practices we need to design to recruit and retain women in the sciences and engineering. One notable feature of this review will be its focus, not simply on women generally (which includes the substantial number of white women who I believe are less influenced by sexism than foreign women and minority women) but on women from foreign lands and women of African ancestry and Hispanic heritage.

Before I begin my presentation of data on the extent of sexism in the student population, I want to make one observation. In my current research I was concerned if somehow graduate schooling in the S and E disciplines inherently reinforced sexism. This might be indicated, perhaps, by students with sexist attitudes feeling more encouraged in their studies than those without such attitudes.

I was quite pleased to find that generally the students—both men and women—who were less sexist in their attitudes, were more likely to feel encouraged in their graduate studies than the students with sexist attitudes—although the differences were not necessarily large.

However, I found one important exception to this broad finding. When I looked at the simple relationship between whether "students felt that the schools offered them opportunities to do research on problems in the field" and "whether they felt men were better professors than women," 81.8 percent of the more sexist students versus 68.9 percent of all others agreed. (Table 4.1a below)

Furthermore, when I looked at the same relationship among men only, the relationship was somewhat stronger: 85.1 percent of the sexist men versus 69.2 percent of the less sexist men felt that the school offered them opportunities to do research on problems in the field. I see this as significant since it shows that more of the sexist students seem to feel they are being prepared for careers as researchers in their discipline than do the less sexist students. I shall discuss this issue more below.

Table 4.1a: School Encourages Students by Providing Opportunities to Work on Research Problems by Men Make Better Professors

	Men Make Better Professors	
Opportunity for Research	Strongly Agree/Agree	All Others
Yes	81.8 (63)	68.9 (739)
No	18.2 (14)	31.1 (334)
Total	77	1,073

Pearson Chi-square=5.109, p=.024 (continuity correction); Fisher's Exact Test p=.02.

Table 4.1b: School Encourages Students by Providing Opportunities to Work on Research Problems by Men Make Better Professors (Men Only)

	Men Make Better Professors	
Opportunity for Research	Strongly Agree/Agree	All Others
Yes	85.1 (40)	69.2 (386)
No	14.9 (7)	30.8 (172)
Total	47	558

Pearson Chi-square=4.544, p=.033 (continuity correction); Fisher's Exact Test p=.02.

The Problem of Sexism in the S and E Student Population—Introductory Remarks

If we are to be successful in increasing the number of women and minorities in the S and E student body and faculty, we must create a welcoming climate. A necessary part of that campaign to make women and minorities welcome is to combat sexism among the student population.

Sexism's effects are seriously damaging: women who feel that men make better scientists or better professors, I believe, are less able to resist pressure from spouses, or in-laws to give up their aspirations to be scientists and engineers. They are less able to resist entreaties to content themselves either with low-end positions in their fields or with being stay at home mothers. Men who believe that men make better scientists are less likely to have women team members, less likely to vote for tenure for women scientists or promotions for women when they know they are assessing a woman.

Sexism manifests itself in different ways in different contexts. For example, in graduate school, a sexist atmosphere can exist when fellow students refuse to invite women students to study with them for the comprehensive examinations, or when professors do not call on women to answer questions in class. It is not necessary that women students in the S and E disciplines are expected to fetch coffee for their fellow students or take notes for them in class for a sexist atmosphere to exist. And a professor can create a sexist atmosphere when, as Ben Barres recalls, a professor told him (when he was still Barbara) that he (then a she) couldn't have solved a difficult homework problem that no other student in the class was able to solve so Barbara's boyfriend must have solved it for her. (Begley, 2006)

Once the women begin their careers in the scientific research environment, the women's own feelings (and those of others), that women do not make as good professors as men, or do not make as good scientists as men, or do not make good team members, might have negative consequences for the women's careers. For example, research has shown, "When evaluators rated writing skills, resumes, journal articles, and career paths, they gave lower ratings on average if they were told . . . the subject of evaluation was a woman (Handelsman et al., 2005, p. 1191). In turn, this could result in not granting tenure to a woman researcher in the department whose accomplishments merit tenure. Or the employer or someone acting on the employer's behalf who feels that women do not make as good scientists as men might not hire a woman researcher with a doctoral degree if a man is available even if the woman's credentials on paper are superior. Or the scientist or engineer who feels that women do not make as good team members as men might not be willing to work on a team that included women investigators.

I started to explore the problem of sexist attitudes in CATSG showing that compared to other groups substantially more Asian male students felt men make better professors (about 23 percent), better scientists (about 28 percent), and better team members (14 percent) than women; among white male students the scores were five percent, nine percent, and five percent respectively (see Tables 3.1a-c in CATSG). However, I decided to wait until I had collected data on Hispanic heritage and African ancestry students before I investigated this issue in the greater depth that I believe it deserves.

ASSESSING SEXISM IN THE S AND E STUDENT POPULATION: THE RESEARCH QUESTIONS

In this chapter I look at sexism as indicated by student responses to several questions:

- Do you believe that men make better professors? Better scientists and engineers? Better team members?
- Do you believe that your classmates think men make better professors? Better scientists and engineers? Better team members?

The specific intent of this chapter is twofold: first, I want to estimate the amount of sexism in the student population. I do this primarily by looking at the ethnic distribution of sexism, the disciplinary distribution, and the distribution within different atmospheres of respect for gender differences and respect for ethnic differences.

Second I want to examine the effect of sexism on whether students feel encouraged by the school, e.g., professors informally encourage students to start or continue research. While most of the results shown will be intuitively obvious, a few surprises did appear in the data. Generally, I offer possible explanations for these unexpected findings—however, in some cases I confess that I am not satisfied with my suggested explanations.

ASSESSING STUDENT SEXISM: RESEARCH INDICATORS AND MEASURED INCIDENCE

I measured sexism in the science and engineering population with six indicators:

- Men make better professors
- Men make better scientists and engineers
- Men make better team members
- Classmates think men make better professors
- Classmates think men make better scientists and engineers
- Classmates think men make better team members

Each of these indicators has advantages and disadvantages as a sole measure of sexism. However, regardless of the indicator chosen the picture they present is that sexism is a problem in the student body.

A word about the tabular data presentation is in order. I chose tables based on two criteria: statistical significance of the results based on the probability of a chance result being less than five percent and "effect size" being ten percent difference or greater on the policy variable depending on conditions in the predictor or explanatory variable. I made one exception in the case of the latter criterion by showing a table with a difference between scores of the two groups of just below ten percent because that difference met the statistical criterion of .05.

Another point is necessary to mention here. Many of the analyses refer to differences between group A and group B students. This is a short hand expression I have adopted to refer to the students who rated their professors average to below average on respect for gender differences or average to below average on respect for ethnic differences (group A) or to students who rated their professors as high on respect for gender differences or high on respect for ethnic differences (group B).

One final housekeeping detail is relevant here. I do not show every table about whose results I may speak because that would be very bulky and costly. (If a reader is interested in certain tables not shown, I encourage him or her to correspond with me directly). The primary reason for my organizing the presentation this way is to make the argument as succinct as possible.

Before I turn to the analyses, I want to make two important general points: First, the school atmosphere is partly a result of attitudes students bring with them—attitudes the students have picked up previously, including from prior teachers, playmates, and parents—and partly a result of attitudes the schools foster. The attitudes the students bring with them are patterned: i.e., they show some constancy across individuals with certain similar traits such as the same ethnic background, or gender etc. With regard to sexist attitudes, (e.g., men make better professors, scientists, team members) few white students and African ancestry students who arrive on campus come with these attitudes compared to Asian students and even Hispanic heritage students.

The second point I want to make in this chapter is that school policies can make a significant difference. If a school's policy is welcoming, that will be reflected in the low incidence of sexist attitudes among students.

On the other hand, I must point out a formal school policy proscribing sexist behavior is not enough to demonstrate that the school has a welcoming atmosphere. I used "professors respect for gender differences" as a proxy for a formal school policy proscribing sexism" and found that this variable made little difference in students' sexist attitudes. Much more important in understanding whether the students had sexist attitudes were students' ethnic backgrounds and their perceptions that the school encouraged them specifically (e.g., by not burdening them with undergraduate teaching responsibilities).

What the data are telling me, I maintain, is that a school committed to educating its students, committed to seeing them succeed in their studies—as evidenced by their financial aid policies, their efforts to bring in speakers to talk about opportunities in the field, etc.—will be much more successful in teaching the students respect for one another and for themselves while discouraging sexist and racist attitudes than a school that issues pronouncements about its dismay over sexism while demonstrating little interest in the students' graduate school experience.

Incidence of Sexism in the S and E Graduate Student Population

I now want to turn to the results of my survey, beginning with some data on the extent of sexism and then looking at some issues that need further consideration.

I estimate that between five and ten percent of the S and E graduate student population have sexist attitudes. The range reflects the differing results obtained using the six indicators. Overall, the problem of sexism in the student population is not a major problem from the standpoint of numbers of students exhibiting these attitudes. Yet, this is not ground for complacency. Sexism can easily become a major problem in a department, or even in an entire school. For example, unless strong countermeasures are taken concomitantly, recruiting large numbers of sexist students from societies where women traditionally have been expected to be deferential to men—China, Japan, Middle Eastern societies—will lead to a serious problem of sexism on campus. Weak administrative support for recruiting women faculty in traditionally very masculine fields such as electrical engineering or physics will allow sexism to flourish at least in such very male dominated departments. Limited recruitment of senior women faculty in sciences and engineering likewise will fan the flames of sexism. In short, sexism is a problem that requires constant vigilance to control.

It is now time to turn to a detailed presentation of the results of my inquiry. I will begin the analysis with data on the stimulus "men make better professors" and will also include results for my analysis of "classmates think men make better professors." Then I will present analyses for "men make better scientists" and "classmates think men make better scientists" and finish with results for "men make better team members" and "classmates think men make better team members."

Table 4.2a-f shows a distribution of graduate students in the S and E disciplines by ethnic/gender group similar to the tables 3.1a-3.1f in CATSG where I first discussed sexism. The main differences between the earlier tables and these new ones are that now I include African ancestry males and Hispanic heritage males as separate categories, and of course, the numbers are greater than in the tables presented in CATSG. On the indicator, "men make better scientists and engineers," more than a quarter of the Asian students strongly agreed or agreed. Even on other indicators, as many as 22 percent strongly agreed or agreed and no less than ten percent agreed strongly or agreed on the most conservative of the measures.

Table 4.2a: Men Make Better Professors by Student Background

Student Background	Strongly Agree/Agree	All Others	Total
White Male	3.0 (9)	97.0 (290)	299
White Male: CO	1.9 (1)	98.1 (52)	53
White Male: NYC	1.8 (2)	98.2 (112)	114
White Male: NC	4.6 (6)	95.4 (125)	131
Asian Male	20.3 (32)	79.7 (126)	158
Black Male	0.0 (0)	100.0 (32)	32
Black Male: Native	0.0 (0)	100.0 (19)	19
Black Male: Foreign	0.0 (0)	100.0 (13)	13
Hispanic Male	5.2 (6)	94.8 (110)	116
Female	5.5 (30)	94.5 (515)	545
Total	77	1073	1150

Pearson chi-square test: Chi-square=56.952, p<.000, df=4

Table 4.2b: Men Make Better Scientists and Engineers by Student Background

Student Background	Strongly Agree/Agree	All Others	Total
White Male	7.7 (23)	92.3 (276)	299
White Male: CO	3.8 (2)	96.2 (51)	53
White Male: NYC	5.3 (6)	94.7 (108)	114
White Male: NC	11.5 (15)	88.5 (116)	131
Asian Male	25.9 (41)	74.1 (117)	158
Black Male	3.1 (1)	96.9 (31)	32
Black Male: Native	0.0 (0)	100.0 (19)	19
Black Male: Foreign	7.7 (1)	92.3 (12)	13
Hispanic Male	4.3 (5)	95.7 (111)	116
Female	4.2 (23)	95.8 (522)	545
Total	93	1057	1150

Pearson chi-square test: Chi-square=82.134, p<.000, df=4

Table 4.2c: Men Make Better Team Members by Student Background

Student Background	Strongly Agree/Agree	All Others	Total
White Male	4.3 (13)	95.7 (286)	299
White Male: CO	3.8 (2)	96.2 (51)	53
White Male: NYC	4.4 (5)	95.6 (109)	114
White Male: NC	4.6 (6)	95.4 (125)	131
Asian Male	13.3 (21)	86.7 (137)	158
Black Male	3.1 (1)	96.9 (31)	32
Black Male: Native	5.3 (1)	94.7 (18)	19
Black Male: Foreign	0.0 (0)	100.0 (13)	13
Hispanic Male	7.8 (9)	92.2 (107)	116
Female	6.1 (33)	93.9 (512)	545
Total	77	1073	1150

Pearson chi-square test: Chi-square=14.861, p=.005, df=4

Table 4.2d: Classmates Think Men Make Better Professors by Student Background

Student Background	Strongly Agree/Agree	All Others	Total
White Male	6.0 (18)	94.0 (281)	299
White Male: CO	5.7 (3)	94.3 (50)	53
White Male: NYC	4.4 (5)	95.6 (109)	114
White Male: NC	6.9 (9)	93.1 (122)	131
Asian Male	17.7 (28)	82.3 (130)	158
Black Male	6.3 (2)	93.8 (30)	32
Black Male: Native	5.3 (1)	94.7 (18)	19
Black Male: Foreign	7.7 (1)	92.3 (12)	13
Hispanic Male	8.6 (10)	91.4 (106)	116
Female	12.3 (67)	87.7 (478)	545
Total	125	1025	1150

Pearson chi-square test: Chi-square=17.366, p=.002, df=4

Table 4.2e: Classmates Think Men Make Better Scientists and Engineers by Student Background

Student Background	Strongly Agree/Agree	All Others	Total
White Male	7.7 (23)	92.3 (276)	299
White Male: CO	7.5 (4)	92.5 (49)	53
White Male: NYC	7.0 (8)	93.0 (106)	114
White Male: NC	8.4 (11)	91.6 (120)	131
Asian Male	19.0 (30)	81.0 (128)	158
Black Male	12.5 (4)	87.5 (28)	32
Black Male: Native	5.3 (1)	94.7 (18)	19
Black Male: Foreign	23.1 (3)	76.9 (10)	13
Hispanic Male	6.0 (7)	94.0 (109)	116
Female	15.2 (83)	84.8 (462)	545
Total	147	1003	1150

Pearson chi-square test: Chi-square=20.072, p<.000, df=4

Table 4.2f. Classmates Think Men Make Better Team Members by Student Background

Student Background	Strongly Agree/Agree	All Others	Total
White Male	3.7 (11)	96.3 (288)	299
White Male: CO	3.8 (2)	96.2 (51)	53
White Male: NYC	4.4 (5)	95.6 (109)	114
White Male: NC	3.1 (4)	96.9 (127)	131
Asian Male	12.0 (19)	88.0 (139)	158
Black Male	6.3 (2)	93.7 (30)	32
Black Male: Native	0.0 (0)	100.0 (19)	19
Black Male: Foreign	15.4 (2)	84.6 (11)	13
Hispanic Male	7.8 (9)	92.2 (107)	116
Female	9.7 (53)	90.3 (492)	545
Total	94	1056	1150

Pearson chi-square test: Chi-square=13.102, p=.011, df=4

Among other student component populations, results were much lower and overall the results show somewhere between five and ten percent of the total graduate student population exhibit sexist attitudes. Nevertheless, it would be foolish to draw the conclusion that since the overall rate of sexism suggested by the tables is low, sexism is not a serious problem for graduate schools. Consider the headline of an article in the Wall Street Journal, April 3, 2012 "Chinese Applicants Flood U.S. Graduate Schools." According to the article an estimated "514,298 applications were sent to U.S. graduate schools by prospective international students for fall 2012," representing an increase of nine percent from 2011. Furthermore, the Council of Graduate Schools, a trade association quoted in the article, stated, "International students comprise about 14 % of students at U.S. graduate schools." However, this percentage seems likely to grow. The article points out that "

> Ohio State University, with more than 10,000 graduate students, opened a 'gateway office' in Shanghai in 2010 to coordinate recruiting, alumni activities and fundraising efforts. While international applications rose 18% for all the school's graduate programs this year, they jumped 29% from China.

Ohio State University's plan is deeply troubling on two counts. First, it may be creating a worse problem of sexism in the S and E departments not only at Ohio State but any other schools that emulate its recruiting plans since perhaps as many as a third of Asian students arriving on campus come with sexist attitudes. Second, and equally important, we may be worsening our competitive position economically down the line by failing to train enough Americans, including women and minorities, in key disciplines—the ones needed to support technological innovation in the future—that underpins our prosperity. And of course we are doing this while training scientists who once they graduate almost certainly will work for foreign competitors or military industrial complex of potentially hostile powers. Welcoming this flood of foreign students is a short-term fix for money hungry graduate schools that may have dire implications for the United States over the next twenty years or so.

Table 4.3a-f examines the same relationship as Tables 4.2a-f except that the former present results only for students within the hard sciences. What is most telling in the tables is not just that the proportion of Asian male graduate students scoring high on sexist attitudes is higher than any other group. This is perturbing but not unexpected in light of the results found both in the CATSG and in Tables 4.2a-f. In some of the tables, I noticed that in other ethnic/racial groups where I had seen little evidence of sexism in Tables 4.2a-f. I found the percentage scoring high on sexism was in double digits. For example, in the Table 4.2d, in response to the query "Do you agree that classmates think men make better professors?" 17.1 percent of the Asian male students felt their classmates thought that men make better professors than women. However, 15.7 percent of the women also felt their classmates thought men make better professors than women.

Table 4.3a: Men Make Better Professors by Student Background for Hard Sciences Students

Student Background	Strongly Agree/Agree	All Others	Total
White Male	3.2 (7)	96.8 (215)	222
Asian Male	22.9 (32)	77.1 (108)	140
Black Male	0.0 (0)	100.0 (26)	26
Hispanic Male	4.9 (4)	95.1 (77)	81
Female	7.3 (22)	92.7 (278)	300
Total	65	704	769

Pearson chi-square test: Chi-square=49.776, p<.000, df=4

Table 4.3b: Men Make Better Scientists and Engineers by Student Background for Hard Sciences Students

Student Background	Strongly Agree/Agree	All Others	Total
White Male	8.6 (19)	91.4 (203)	222
Asian Male	27.9 (39)	72.1 (101)	140
Black Male	3.8 (1)	96.2 (25)	26
Hispanic Male	6.2 (5)	93.8 (76)	81
Female	6.0 (18)	94.0 (282)	300
Total	82	687	769

Pearson chi-square test: Chi-square=54.310, p<.000, df=4

Table 4.3c: Men Make Better Team Members by Student Background for Hard Sciences Students

Student Background	Strongly Agree/Agree	All Others	Total
White Male	5.0 (11)	95.0 (211)	222
Asian Male	14.3 (20)	85.7 (120)	140
Black Male	3.8 (1)	96.2 (25)	26
Hispanic Male	7.4 (6)	92.6 (75)	81
Female	6.3 (19)	93.7 (281)	300
Total	57	712	769

Pearson chi-square test: Chi-square=12.582, p=.014, df=4

Table 4.3d: Classmates Think Men Make Better Professors by Student Background for Hard Sciences Students

Student Background	Strongly Agree/Agree	All Others	Total
White Male	6.8 (15)	93.2 (207)	222
Asian Male	17.1 (24)	82.9 (116)	140
Black Male	7.7 (2)	92.3 (24)	26
Hispanic Male	8.6 (7)	91.4 (74)	81
Female	15.7 (47)	84.3 (253)	300
Total	95	674	769

Pearson chi-square test: Chi-square=13.982, p=.007, df=4

We can debate whether women have internalized this perception that men make better professors. What we cannot debate is that a substantial number of women students in the hard sciences perceive their classmates thinking this is so. And given that they have few women role models in the hard sciences—perhaps also, as Etzkowitz et al. noticed, somewhat isolated from other students—they may be vulnerable to pressures to leave, especially when they encounter setbacks in their quest for the doctorate, which frequently happens when students undertake an arduous course of study.

Table 4.3e: Classmates Think Men Make Better Scientists and Engineers by Student Background for Hard Sciences Students

Student Background	Strongly Agree/Agree	All Others	Total
White Male	9.5 (21)	90.5 (201)	222
Asian Male	17.9 (25)	82.1 (115)	140
Black Male	15.4 (4)	84.6 (22)	26
Hispanic Male	7.4 (6)	92.6 (75)	81
Female	18.7 (56)	81.3 (244)	300
Total	112	657	769

Pearson chi-square test: Chi-square=13.275, p=.010, df=4

In Table 4.3e, where students are asked if they agree that their classmates think that men make better scientists and engineers, we see 17.9 percent of the Asian male students agreeing or strongly agreeing that this is true. However, we also see 18.7 percent of the women students a strongly agreeing or agreeing this is true; 15.4 percent of the African ancestry male students strongly agreeing or agreeing that this is true; and 9.5 percent of the white male students strongly agreeing or agreeing that this is true. Can that many of each group feel this way about their fellow students if the amount of sexism among the students is minute?

Table 4.3f: Classmates Think Men Make Better Team Members by Student Background for Hard Sciences Students

Student Background	Strongly Agree/Agree	All Others	Total
White Male	4.5 (10)	95.5 (212)	222
Asian Male	11.4 (16)	88.6 (124)	140
Black Male	7.7 (2)	92.3 (24)	26
Hispanic Male	7.4 (6)	92.6 (75)	81
Female	11.3 (34)	88.7 (266)	300
Total	68	701	769

Pearson chi-square test: Chi-square=8.903, p=.064, df=; Cramer's V=.108

Because in the hard sciences women's penetration of the top echelons has been limited by comparison to other fields, and feminist writers such as Cockburn and Wajcman have focused on the hard sciences and engineering, it is understandable that we would inquire if sexism is indeed a problem in those disciplines. However, it is wrong to ignore the possibility that the soft sciences also have a problem with sexism. As is evident in Table 4.4a-c, I found that among soft sciences students, 11.8 percent of Asian male students believed that men make better scientists and engineers.

Table 4.4a: Men Make Better Professors by Student Background for Soft Sciences Students

Student Background	Strongly Agree/Agree	All Others	Total
White Male	2.6 (2)	97.4 (74)	76
Asian Male	0 (0)	100.0 (17)	17
Black Male	0 (0)	100.0 (6)	6
Hispanic Male	5.7 (2)	94.3 (33)	35
Female	3.3 (8)	96.7 (235)	243
Total	12	365	377

Pearson chi-square test: Chi-square=1.568, p=.814, df=4

Table 4.4b: Men Make Better Scientists and Engineers by Student Background for Soft Sciences Students

Student Background	Strongly Agree/Agree	All Others	Total
White Male	5.3 (4)	94.7 (72)	76
Asian Male	11.8 (2)	88.2 (15)	17
Black Male	0.0 (0)	100.0 (6)	6
Hispanic Male	0.0 (0)	100.0 (35)	35
Female	2.1 (5)	97.9 (238)	243
Total	11	366	377

Pearson chi-square test: Chi-square=8.040, p=.090, df=4
Cramer's V=.146

Table 4.4c: Men Make Better Team Members by Student Background for Soft Sciences Students

Student Background	Strongly Agree/Agree	All Others	Total
White Male	2.6 (2)	97.4 (74)	76
Asian Male	5.9 (1)	94.1 (16)	17
Black Male	0 (0)	100.0 (6)	6
Hispanic Male	8.6 (3)	91.4 (32)	35
Female	5.8 (14)	94.2 (229)	243
Total	20	357	377

Pearson chi-square test: Chi-square=2.273, p=.686, df=4

Perhaps more intriguing was the results for soft sciences respondents when I looked at students' perceptions of what their classmates believed. Table 4.4d shows that among soft sciences respondents 23.5 percent of the Asian male students strongly agreed or agreed that their classmates believed better professors as did 8.6 of the Hispanic male students and 8.2 percent of the women students. (See Table 4.4e below).

When queried as to whether their classmates felt that men make better scientists and engineers, 29.4 percent of the Asian male students in the soft sciences and 11.1 percent of the women students perceived that their classmates felt this way. (See Table 4.4e below).

Finally, when I asked respondents, "Do men make better team members?" 17.6 percent of the Asian male respondents strongly agreed or agreed. But 8.6 percent of the Hispanic males also strongly agreed or agreed, as did 7.8 percent of the female respondents (see Table 4.4f below).

Table 4.4d: Classmates Think Men Make Better Professors by Student Background for Soft Sciences Students

Student Background	Strongly Agree/Agree	All Others	Total
White Male	3.9 (3)	96.1 (73)	76
Asian Male	23.5 (4)	76.5 (13)	17
Black Male	0 (0)	100.0 (6)	6
Hispanic Male	8.6 (3)	91.4 (32)	35
Female	8.2 (20)	91.8 (223)	243
Total	30	347	377

Pearson chi-square test: Chi-square=7.858, p=.097, df=4; Cramer's V=.144

Table 4.4e: Classmates Think Men Make Better Scientists and Engineers by Student Background for Soft Sciences Students

Student Background	Strongly Agree/Agree	All Others	Total
White Male	2.6 (2)	97.4 (74)	76
Asian Male	29.4 (5)	70.6 (12)	17
Black Male	0 (0)	100.0 (6)	6
Hispanic Male	2.9 (1)	97.1 (34)	35
Female	11.1 (27)	88.9 (216)	243
Total	35	342	377

Pearson chi-square test: Chi-square=15.465, p=.004, df=4; Cramer's V=.203

Table 4.4f: Classmates Think Men Make Better Team Members by Student Background for Soft Sciences Students

Student Background	Strongly Agree/Agree	All Others	Total
White Male	1.3 (1)	98.7 (75)	76
Asian Male	17.6 (3)	82.4 (14)	17
Black Male	0 (0)	100.0 (6)	6
Hispanic Male	8.6 (3)	91.4 (32)	35
Female	7.8 (19)	92.2 (224)	243
Total	26	351	377

Pearson chi-square test: Chi-square=7.666, p=.105, df=4; Cramer's V=.143

Pointing to women's relatively insignificant presence in the faculties of physics and electrical engineering departments, feminist scholars have long charged that women in the hard sciences face considerable sexism (see for example, Cockburn and Wajcman on engineering); however, as the above data demonstrate sexism also is present in the soft sciences (as has been shown for example by Ehrenreich and English, 1973). These data demonstrate that more important in explaining sexism in a discipline than the characteristics of the discipline (e.g., that spatial reasoning skills are a prerequisite for doing much of the research in the field) itself is the simple fact that hardly any women faculty are found in the discipline. The absence or near absence of women among the faculty in the department cements student views that this is somehow the way things should be.

Social psychological research, especially the famed "Robbers' Cave Experiment" (Sherif et al., 1988), forcefully demonstrates that universities can effectively fight sexism if they choose to do so. We do not have to produce scientists and engineers whose beliefs include the perception that women are not as good professors as men or not as good scientists and engineers or not as effective team members in research projects.

However, as schools come to depend on these foreign students for the revenue they bring in, an urgent question arises as to how strenuously will the schools seek to combat sexism or racism to make the school climate more welcoming for women and African ancestry and Hispanic heritage students? We must not let the universities' hunger for money to meet their fiscal needs cause them to backslide on progress they are making in opening the doors to women and minorities. But this will be a difficult objective to meet. The universities are not simply interested in recruiting more foreign students (desirable because they bring in tuition fees higher than in-state residents have to pay). The universities are planning to step up recruiting in countries with long histories of expecting women to be subordinate and deferential to male authority figures. China, for example, once bound the feet of women and even now news filters out of female infanticide in the countryside—an unfortunate consequence of China's one child per family rule in a nation where parents have traditionally preferred male children. We need to press the school administrations for answers to the question: How will the university assure that women, and African ancestry and Hispanic heritage students feel welcome while the schools are busy recruiting foreign students from Asia who are willing to pay high tuition fees?

If the universities are reluctant to address sexism on campus for fear of offending tuition paying Asian students it is surely not because tools for combating sexism are unavailable. The fact is, universities can avail themselves of effective tools for combating sexism although not all universities seem to to do this with appropriate vigor as demonstrated in CATSG (see pp. 52-61). For example, in schools in the New York City/ suburban New Jersey area, 28.2 percent of the Asian male graduate students strongly agreed or agreed that men make better professors and 29.49 percent felt men make better scientists and engineers. In the schools in the Research Triangle of North Carolina, on the other hand, 12.8 percent of the Asian male students—15.4 percent fewer than in the NY/NJ region—believed that men make better professors and 20.83 percent believed men make better scientists and engineers. (It is possible, of course, that students going to North Carolina schools were more enlightened than those going to New York City/ suburban New Jersey schools but I rather doubt this was the reason for the difference in regional results).

SEXISM OF STUDENTS AS A CONSEQUENCE OF PROFESSORS' PERCEIVED LACK OF RESPECT FOR GENDER DIFFERENCES

With these findings from prior research in mind, I decided to ask the following question: would there be a difference in the proportion of students holding sexist attitudes if the school atmosphere were varied. To test this idea, I evaluated students on the basis of their assessments of professors' respect for gender differences and their respect for ethnic differences. My question then became, Will different proportions of students of each gender and/or ethnic group agree with sexist stimuli depending on whether they gave the professors above average ratings on respect for gender differences (or ethnic differences) versus average or below average ratings on respect for gender differences or ethnic differences?

Overall, I found small statistically significant differences (.05) between the two genders in the expected direction. However, the differences did not meet my effect criterion. I am reluctant to drop my insistence on larger differences between groups (ten percent or greater) than I found but I do think it important to note the results—found in a number of tables—and let the reader decide.

Table 4.5a: Men Make Better Scientists and Engineers by Gender for Students who Rate Professors Average or Below on Respect for Gender Differences

Average or Below Respect for Gender Differences	Strongly Agree/Agree	All Others	Total
Male	12.4 (39)	87.6 (275)	314
Female	4.0 (13)	96.0 (314)	327
Total	52	589	641

Pearson chi-square test: Chi-square=15.325, p=.000, df=1

Table 4.5b: Classmates Think Men Make Better Scientists and Engineers by Gender for Students who Rate Professors Average or Below on Respect for Gender Differences

Average or Below Respect for Gender Differences	Strongly Agree/Agree	All Others	Total
Male	13.7 (43)	86.3 (271)	314
Female	18.3 (60)	81.7 (267)	327
Total	103	538	641

Pearson chi-square test: Chi-square=2.573, p=.109, df=1; Gamma=-.158

However, I found differences within specific ethnic groups that did meet my effect criterion. For example, when Asian graduate students were asked "Do classmates think men make better professors, 23.9 percent of the group A students strongly agreed or agreed whereas only 9.3 percent of the group B students strongly agreed or agreed (see Tables 3.75-76); similarly 27.1 percent of Asian students in group A strongly agreed or agreed that classmates think men make better scientists and engineers versus 12.0 percent of group B Asian students. In respect to team members, although the differences were in the same direction they did not meet my ten percent effect criterion (15.5 percent in group A versus 7.4 percent in group B).

Table 4.6a: Classmates Think Men Make Better Professors by Ethnicity for Students who Rate Professors Average or Below on Respect for Gender Differences

Average or Below Respect for Gender Differences	Strongly Agree/Agree	All Others	Total
White	9.6 (33)	90.4 (309)	342
Asian	23.9 (37)	76.1 (118)	155
Black	16.3 (8)	83.7 (41)	49
Hispanic	10.5 (10)	89.5 (85)	95
Total	88	553	641

Pearson chi-square test: Chi-square=19.370, p<.000, df=3

Table 4.6b: Classmates Think Men Make Better Professors by Ethnicity for Students who Rate Professors Above Average on Respect for Gender Differences

Above Average on Respect for Gender Differences	Strongly Agree/Agree	All Others	Total
White	6.4 (15)	93.6 (218)	233
Asian	9.3 (10)	90.7 (98)	108
Black	0.0 (0)	100.0 (22)	22
Hispanic	9.3 (10)	90.7 (97)	107
Total	35	435	470

Pearson chi-square test: Chi-square=3.819, p=.363, df=3

Table 4.6c: Classmates Think Men Make Better Scientists and Engineers by Ethnicity for Students who Rate Professors Average or Below on Respect for Gender Differences

Average or Below Respect for Gender Differences	Strongly Agree/Agree	All Others	Total
White	10.5 (36)	89.5 (306)	342
Asian	27.1	72.9	155

Average or Below Respect for Gender Differences	Strongly Agree/Agree	All Others	Total
	(42)	(113)	
Black	18.4 (9)	81.6 (40)	49
Hispanic	16.8 (16)	83.2 (79)	95
Total	103	538	641

Pearson chi-square test: Chi-square=22.001, p<.000, df=3

Table 4.6d: Classmates Think Men Make Better Scientists and Engineers by Ethnicity for Students who Rate Professors Above Average on Respect for Gender Differences

Above Average on Respect for Gender Differences	Strongly Agree/Agree	All Others	Total
White	8.2 (19)	91.8 (214)	233
Asian	12.0 (13)	88.0 (95)	108
Black	0.0 (0)	100.0 (22)	22
Hispanic	7.5 (8)	92.5 (99)	107
Total	40	430	470

Pearson chi-square test: Chi-square=3.956, p=.266, df=3

Table 4.6e: Classmates Think Men Make Better Team Members by Ethnicity for Students who Rate Professors Average or Below on Respect for Gender Differences

Average or Below Respect for Gender Differences	Strongly Agree/Agree	All Others	Total
White	7.6 (26)	92.4 (316)	342
Asian	15.5 (24)	84.5 (131)	155
Black	12.2 (6)	87.8 (43)	49
Hispanic	12.6 (12)	87.4 (83)	95
Total	68	576	641

Pearson chi-square test: Chi-square=7.693, p=.053, df=3; Cramer's V=.110

Table 4.6f: Classmates Think Men Make Better Team Members by Ethnicity for Students who Rate Professors Above Average on Respect for Gender Differences

Above Average on Respect for Gender Differences	Strongly Agree/Agree	All Others	Total
White	1.7 (4)	98.3 (229)	233
Asian	7.4 (8)	92.6 (100)	108
Black	0.0	100.0	22

Above Average on Respect for Gender Differences	Strongly Agree/Agree	All Others	Total
	(0)	(22)	
Hispanic	11.2	88.8	107
	(12)	(95)	
Total	24	446	470

Pearson chi-square test: Chi-square=16.128, p=.001, df=3

Asian students were not the only ones to show significant differences between group A and group B students. As Table 4.6a demonstrates, in response to the stimulus "Classmates think men make better professors", 16.3 percent of African ancestry students in group A versus none of the African ancestry students in group B strongly agreed or agreed. Furthermore, 18.4 percent of the African ancestry students in group A strongly agreed or agreed that "classmates think men make better scientists and engineers" (see Table 4.6c) whereas none of the group B African ancestry students agreed.

Finally, in respect to whether the students believed their classmates think men make better team members, 12.2 percent of the group A African ancestry students strongly agreed or agreed compared to none of the group B students.

Interestingly, among African ancestry male students, none of the group A students felt men make better professors. Neither did any of the African ancestry males in group B. In regard to whether men make better scientists, group A African males (6.3 percent) did not differ significantly from group B African males (0.0 percent) and did not differ on whether men make better team members from group B either (0.0percent for both groups).

Among Hispanic heritage males, the picture was similar to African ancestry males. When asked if men make better professors, Hispanic heritage males in group A (4.0 percent) and group B (6.3 percent) did not differ; nor did group A (4.0 percent) differ from group B (4.7 percent). In regard to whether men make better team members ten percent of the Hispanic heritage men in group A agreed or strongly agreed versus 6.3 percent in group B. The difference is not statistically significant; however the fact that ten percent of group A feel that men are better team members still suggests that it is important to have a clear policy of affirming that women and men are equally good scientists, professors, and team members in a graduate school and professors are aware of this policy and implement it.

The Sexism of Women S and E Students: Some Correlates

The next part of this chapter addresses sexism among women students. I have already demonstrated its existence in the tabular data presented above (e.g., Table 4.4f). However in this section I want to explore the subject in more depth.

Women's sexism seems counterintuitive. In Marxist terms, it might be classified as a kind of "false consciousness"—laborers' support for capitalism despite the laboring class being exploited by capitalist employers is a classic example. Daniel Little, (n.d.) points out that:

> Marx himself did not use the phrase 'false consciousness,' but he paid extensive attention to the related concepts of ideology and commodity fetishism. Members of a subordinate class (workers, peasants, serfs) suffer from false consciousness in that their mental representations of the social relations around them systematically conceal or obscure the realities of subordination, exploitation, and domination those relations embody.

In the case of women's sexism the "false consciousness" may arise from what abnormal psychologists refer to as "traumatic bonding" in which "victims attach themselves to their tormentors because they know they will be shamed and stigmatized." This possibility would help explain why my data suggest that women's sexism is most pronounced among women from societies and cultures where women traditionally have been expected to be deferential to men.

The first part of my discussion will be an assessment of the extent of women's sexism in the sample population. Then I will inquire into two possible correlates of the women's sexism: (a) whether those who hold sexist beliefs that men make better professors, scientists, etc., also feel less encouraged by the university and (b) whether the women are more likely to hold these beliefs depending on the school atmosphere; e.g., professors seem to be average or below average in respect for gender differences, hard sciences versus soft, competitive or not.

In this regard, I am not claiming that the correlation of the two kinds of attitudes implies any particular direction of causation. It is plausible that the causation is two-way. A woman student can feel discouraged in her studies and that may lead her to believe that classmates think men make better professors, scientists, or team members. Alternatively, a woman student may sense that classmates think men make better professors and that may lead her to feel discouraged about her chances to have a good career rather than encouraged by the school to succeed.

Table 4.7a: Men Make Better Professors by Ethnicity for Females who Rate Professors Average or Below on Respect for Gender Differences

Average or Below Respect for Gender Differences	Strongly Agree/Agree	All Others	Total
White Female	2.2 (4)	97.8 (181)	185
Asian Female	14.1 (9)	85.9 (55)	64
Black Female	12.1 (4)	87.9 (29)	33
Hispanic Female	2.2 (1)	97.8 (44)	45
Total	18	309	327

Pearson chi-square test: Chi-square=16.694, p=.001, df=3

Table 4.7a shows that a sizable fraction of women graduate students in the Asian female (14.1 percent) and African ancestry female (12.1 percent)

component groups of those students who rated the professors average to below average (on respect for gender differences) felt men make better professors. Interestingly, while African heritage women among those students who rated their professors above average on respect for gender differences unanimously disagreed that men make better professors, 20 percent of the Asian female students agreed or strongly agreed that men made better professors.

Table 4.7b: Men Make Better Professors by Ethnicity for Females who Rate Professors Above Average on Respect for Gender Differences

Above Average on Respect for Gender Differences	Strongly Agree/Agree	All Others	Total
White Female	1.9 (2)	98.1 (104)	106
Asian Female	20.0 (9)	80.0 (36)	45
Black Female	0.0 (0)	100.0 (7)	7
Hispanic Female	2.3 (1)	97.7 (42)	43
Total	12	189	201

Pearson chi-square test: Chi-square=20.389, p<.000, df=3

Table 4.8a: Men Make Better Scientists and Engineers by Ethnicity for Females who Rate Professors Average or Below on Respect for Gender Differences

Average or Below Respect for Gender Differences	Strongly Agree/Agree	All Others	Total
White Female	1.6 (3)	98.4 (182)	185
Asian Female	10.9 (7)	89.1 (57)	64
Black Female	6.1 (2)	93.9 (31)	33
Hispanic Female	2.2 (1)	97.8 (44)	45
Total	13	314	327

Pearson chi-square test: Chi-square=11.549, p=.009, df=3

Table 4.8a shows that among Asian females graduate students a sizable fraction think men make better scientists and engineers. Among students who rated their professors average or below average on respect for gender differences (group A), 10.9 percent of the Asian women students felt men were superior as scientists. And Table 4.8b shows among Asian women students who rated their professors superior on respect for gender differences, 15.6 percent felt men make better scientists and engineers. The differences are too small between the two groups of Asian female students to be regarded as significant but the magnitude of the percentage of Asian women who, regardless of school policy favoring gender diversity among S and E faculty, continue to assert men make better scientists is troubling from a policy standpoint. It also testifies to the enduring influence on Asian women of the sexist views they acquired in their formative years. Consequently, this finding constitutes some of the most compelling evidence of the correctness of Cole and Singer's fundamental point about prior socialization (before graduate school training) molding the views of the women scientists.

Table 4.8b: Men Make Better Scientists and Engineers by Ethnicity for Females who Rate Professors Above Average on Respect for Gender Differences

Above Average on Respect for Gender Differences	Strongly Agree/Agree	All Others	Total
White Female	0.9 (1)	99.1 (105)	106
Asian Female	15.6 (7)	84.4 (38)	45
Black Female	0.0 (0)	100.0 (7)	7
Hispanic Female	4.7 (2)	95.3 (41)	43
Total	10	191	201

Pearson chi-square test: Chi-square=14.676, p=.002, df=3

Are Asian women unique somehow in this regard? Is the socialization of Asian women somehow different and therefore they are less likely to be persuaded that women scientists and engineers can be just as competent professors and researchers as men? The evidence in my data argues regrettably that this may not be true. Women of other ethnic backgrounds may show sexist attitudes as well.

Table 4.9 shows that 12.1 percent of African ancestry women graduate students among students rating their professors average or below on respect for gender differences felt that men make better team members. Fifteen percent (15.6 percent) of Asian women and 11.1 percent of Hispanic heritage women in group A agreed. Just one white female graduate student out of 185 felt the same way according to Table 4.9.

Table 4.9: Men Make Better Team Members by Ethnicity for Females who Rate Professors Average or Below on Respect for Gender Differences

Average or Below Respect for Gender Differences	Strongly Agree/Agree	All Others	Total
White Female	0.5 (1)	99.5 (184)	185
Asian Female	15.6 (10)	84.4 (54)	64
Black Female	12.1 (4)	87.9 (29)	33
Hispanic Female	11.1 (5)	88.9 (40)	45

Average or Below Respect for Gender Differences	Strongly Agree/Agree	All Others	Total
Total	20	307	327

Pearson chi-square test: Chi-square=24.121, p<.000, df=3

Among group B students, however, i.e., those rating their professors above average on respect for gender differences, 6.7 percent of Asian female students, 0.0% of African ancestry and 9.3 percent of Hispanic heritage women felt that men make better team members.

Women scientists' and engineers' attitudes about team members are significant because they address an issue that affects women scientists' productivity: limited numbers of collaborators versus men (See e.g., Bozeman and Corley, 2004). These data suggest that in trying to raise the productivity of women scientists and engineers we must consider not only men's feelings about who is suitable as a collaborator but women's feelings as well. Women's hesitancy about working with other women scientists and engineers is a neglected issue in the literature but clearly it needs to be studied and strategies for overcoming this problem need to be developed because it has serious implications for women scientists' productivity.

Table 4.10 shows that among group A women graduate students in the S and E disciplines, perceptions of sexism of classmates are quite high across the board. Nearly a third of the Asian women students, for example, (32.8 percent) saw classmates as believing men make better scientists and engineers while about a quarter (24.4 percent) of the Hispanic heritage women, 15.2 percent of the African ancestry, and 12.4 percent of the white women students agreed.

The numbers were depressed considerably when group B women, i.e., those who rated their professors above average on respect for gender differences, were surveyed. Fourteen percent of the Hispanic women, 11.1 percent of the Asian women, 8.5 percent of the white women, and none of the African ancestry women believed their classmates felt men make better scientists and engineers.

Table 4.10: Classmates Think Men Make Better Scientists and Engineers by Ethnicity for Females who Rate Professors Average or Below on Respect for Gender Differences

Average or Below Respect for Gender Differences	Strongly Agree/Agree	All Others	Total
White Female	12.4 (23)	87.6 (162)	185
Asian Female	32.8 (21)	67.2 (43)	64
Black Female	15.2 (5)	84.8 (28)	33
Hispanic Female	24.4 (11)	75.6 (34)	45
Total	60	267	327

Pearson chi-square test: Chi-square=14.600, p=.002, df=3

Finally, among group B women, those rating their professors above average on respect for gender differences, I found that 18.6 percent of the Hispanic women saw their classmates as believing that men make better team members whereas just 1.9 percent of white women, 6.7 percent of Asian women, and none of the African ancestry women felt this way about their classmates' attitudes.

Table 4.11: Classmates Think Men Make Better Team Members by Ethnicity for Females who Rate Professors Above Average on Respect for Gender Differences

Above Average on Respect for Gender Differences	Strongly Agree/Agree	All Others	Total
White Female	1.9 (2)	98.1 (104)	106
Asian Female	6.7 (3)	93.3 (42)	45
Black Female	0.0 (0)	100.0 (7)	7
Hispanic Female	18.6 (8)	81.4 (35)	43
Total	13	188	201

Pearson chi-square test: Chi-square=14.635, p=.002, df=3

These findings from Table 4.10-11 are sobering. One important implication is that the graduate schools cannot entirely overcome prior socialization in implementing a program to diversify their faculty gender composition. Students arrive on campus with attitudes about women that may be difficult to overcome, especially if they grew up in societies where women were expected to be deferential to men. Asian societies in many cases are examples of societies where women historically have been subservient. In Saudi Arabia even now, and perhaps in other conservative Moslem nations in the recent past if not currently, women do not drive, they do not vote, they do not attend school in classes with males etc. In China, female infanticide is still practiced in the rural parts of the country although officially banned. The machismo culture is still a feature of Latin American culture.

The sexism of women themselves must be reckoned in any comprehensive assault on the problem of sexism among graduate students in the S and E disciplines. And unless we make a concerted effort to stamp out sexism in the sciences and engineering we risk grave damage to our pool of competent research scientists and engineers, those most likely to make the important discoveries and innovations in the future that will keep our nation competitive and militarily secure.

In the next section, I want to examine the problem of sexism in the female S and E population in finer detail. I will begin with findings of sexism among women in the hard and soft sciences respectively.

Table 4.12a shows that in the hard sciences, 14.3 percent of Asian female graduate students who rated their professors average or below on respect for gender differences felt men make better professors. In addition, 17.4 percent of

African ancestry women agreed or strongly agreed that men make better professors in the group that rated professors average or below on respect for gender differences. Identical proportions of Asian women and African ancestry women agreed or strongly agreed that men make better team members. (See Table 4.12b)

Table 4.12a: Men Make Better Professors by Ethnicity for Female Hard Science Students who Rate Professors Average or Below on Respect for Gender Differences

Average or Below Respect for Gender Differences	Strongly Agree/Agree	All Others	Total
White Female	3.2 (3)	96.8 (90)	93
Asian Female	14.3 (6)	85.7 (36)	42
Black Female	17.4 (4)	82.6 (19)	23
Hispanic Female	5.0 (1)	95.0 (19)	20
Total	14	164	178

Pearson chi-square test: Chi-square=8.258, p=.041, df=3

Table 4.12b: Men Make Better Team Members by Ethnicity for Female Hard Science Students who Rate Professors Average or Below on Respect for Gender Differences

Average or Below Respect for Gender Differences	Strongly Agree/Agree	All Others	Total
White Female	1.1 (1)	98.9 (92)	93
Asian Female	14.3 (6)	85.7 (36)	42
Black Female	17.4 (4)	85.7 (36)	23
Hispanic Female	15.0 (3)	85.0 (17)	20
Total	14	164	178

Pearson chi-square test: Chi-square=12.591, p=.006, df=3

When I asked respondents what they thought their classmates felt, 38.1 percent of the Asian female graduate students in the hard sciences who rated their professors average or below on respect for gender differences (Group A) asserted that their classmates believed men make better scientists and engineers, 21.7 percent of African ancestry women group A students likewise believed this, 25.0 percent of the group A Hispanic heritage women students agreed or strongly agreed that their classmates felt this way and even 17.2 percent of the Group A white females agreed or strongly agreed that this was the case.

Table 4.13: Classmates Think Men Make Better Scientists and Engineers by Ethnicity for Female Hard Science Students who Rate Professors Average or Below on Respect for Gender Differences

Average or Below Respect for Gender Differences	Strongly Agree/Agree	All Others	Total
White Female	17.2 (16)	82.8 (77)	93
Asian Female	38.1 (16)	61.9 (26)	42
Black Female	21.7 (5)	78.3 (18)	23
Hispanic Female	25.0 (5)	75.0 (15)	20
Total	42	136	178

Pearson chi-square test: Chi-square=7.071, p=.070, df=3; Cramer's V=.199

The toll taken by sexism I maintain is revealed in these data. Large numbers of talented women in the hard sciences, whose faculty instructors (mostly if not entirely male) seem at best lukewarm to the idea of women in the discipline, become convinced that men are better scientists, or better professors, or better team members. Is it plausible that given these attitudes one consequence is that these women avoid working with other women in the field, or come to accept that other women and other men colleagues will not want to collaborate with them? I believe that to be the case. And I further believe that I am hardly the first to detect the internalized sexism of women in the sciences and its destructive effects. Cole and Singer, and later Etzkowitz et al. noted the isolation, the self doubt, the sense of worthlessness felt by women in the sciences in their studies done years earlier than my own. Sexism among women graduate students in the sciences, often acquired in the women's childhood but reinforced by an unwelcoming atmosphere in the graduate school, is an enduring problem and one that will be difficult to combat.

Is sexism among women students uniquely a feature of the hard sciences and especially of those departments where the faculty is not determined to roll out the welcome mat for talented women scientists and engineers? To find out, I next looked at the soft sciences, regarded as more welcoming in general than the hard sciences in the writings of feminist scholars such as Wajcman (1995; also see Fisher, 2007).

Table 4.14a shows that 14.3 percent of Asian women agree or strongly agree that men make better professors among women in the soft sciences who rated their professors as average or below on respect for gender differences (hereafter Group A component). About ten percent (9.5 percent) of the Asian women felt men make better scientists according to Table 4.14b. Perhaps most interesting of all is that 19 percent of the Asian women in the Group A soft sciences cohort agreed or strongly agreed that men make better team members while 8.0 percent of the Hispanic heritage women also agreed or strongly agreed (see Table 4.14c).

Table 4.14a: Men Make Better Professors by Ethnicity for Female Soft Science Students who Rate Professors Average or Below on Respect for Gender Differences

Average or Below Respect for Gender Differences	Strongly Agree/Agree	All Others	Total
White Female	1.1 (1)	98.9 (90)	91
Asian Female	14.3 (3)	85.7 (18)	21
Black Female	0.0 (0)	100.0 (10)	10
Hispanic Female	0.0 (0)	100.0 (25)	25
Total	4	143	147

Pearson chi-square test: Chi-square=12.494, p=.006, df=3

Table 4.14b: Men Make Better Scientists and Engineers by Ethnicity for Female Soft Science Students who Rate Professors Average or Below on Respect for Gender Differences

Average or Below Respect for Gender Differences	Strongly Agree/Agree	All Others	Total
White Female	0.0 (0)	100.0 (91)	91
Asian Female	9.5 (2)	90.5 (19)	21
Black Female	0.0 (0)	100.0 (10)	10
Hispanic Female	0.0 (0)	100.0 (25)	25
Total	2	145	147

Pearson chi-square test: Chi-square=12.166, p=.007, df=3

Table 4.14c: Men Make Better Team Members by Ethnicity for Female Soft Science Students who Rate Professors Average or Below on Respectful of Gender Differences

Average or Below Respect for Gender Differences	Strongly Agree/Agree	All Others	Total
White Female	0.0 (0)	100.0 (91)	91
Asian Female	19.0 (4)	81.0 (17)	21
Black Female	0.0 (0)	100.0 (10)	10
Hispanic Female	8.0 (2)	92.0 (23)	25
Total	6	141	147

Pearson chi-square test: Chi-square=17.292, p=.001, df=3

Table 4.15 shows that in the Group A soft sciences component population, nearly a quarter of the Asian female students (23.8 percent) and the Hispanic heritage female students (24 percent) felt their classmates think men make better scientists and engineers. None of the African ancestry women but 7.7 percent of the white women students in this component population agreed or strongly agreed that their classmates believed men make better scientists and engineers.

Table 4.15: Classmates Think Men Make Better Scientists and Engineers by Ethnicity for Female Soft Science Students who Rate Professors Average or Below on Respect for Gender Differences

Average or Below Respect for Gender Differences	Strongly Agree/Agree	All Others	Total
White Female	7.7 (7)	92.3 (84)	91
Asian Female	23.8 (5)	76.2 (16)	21
Black Female	0.0 (0)	100.0 (10)	10
Hispanic Female	24.0 (6)	76.0 (19)	25
Total	18	129	147

Pearson chi-square test: Chi-square=8.979, p=.030, df=3

The evidence suggests that the type of discipline may matter much less—if at all—than the prior socialization of the women in explaining their sexist views. Female graduate students raised in societies where women have been traditionally expected to be deferential to men, e.g., many conservative Middle Eastern societies, societies in the Orient such as China, and Hispanic societies, arrive in graduate school with views that make them vulnerable to feelings of self doubt and worthlessness. Unless the school reaches out to these women students to involve them they are more likely to be isolated from their peers and from the informal coteries around professors that result in potential collaborators in the future.

While type of discipline may not matter much if my data are corroborated in future studies, I do believe that organizational context can make a difference in sexist attitudes. A friendly school atmosphere, for example, can mitigate some of the inevitable feelings of disappointment that accompany the students' working towards their doctoral degree. A snobbish school atmosphere can exacerbate the sense that the struggle to succeed in getting the doctorate may not be worth the effort required.

Table 4.16a-c shows the results for women who rated their school atmosphere as "snobbish" and their professors as "average or below" on respect for gender differences. Thirty percent of the African ancestry women stated that they agreed or strongly agreed that men make better professors and 20 percent of the Asian women likewise agreed or strongly agreed with this stimulus.

In Table 4.16b, we see that 13 percent of the Asian women and ten percent of the African ancestry women agree that men make better scientists and engineers. Finally Table 4.16c shows that 20 percent of the Asian and African ancestry women respectively felt men make better team members on research projects than women and 8.3 percent of the Hispanic women agreed or strongly agreed.

Table 4.16a: Men Make Better Professors by Ethnicity for Females in Snobbish Atmospheres who Rate Professors Average or Below on Respect for Gender Differences

Average or Below Respect for Gender Differences	Strongly Agree/Agree	All Others	Total
White Female	2.4 (1)	97.6 (40)	41
Asian Female	20.0 (3)	80.0 (12)	15
Black Female	30.0 (3)	70.0 (7)	10
Hispanic Female	0.0 (0)	100.0 (12)	12
Total	7	71	78

Pearson chi-square test: Chi-square=10.971, p=.012, df=3

Table 4.16b: Men Make Better Scientists and Engineers by Ethnicity for Females in Snobbish Atmospheres who Rate Professors Average or Below on Respect for Gender Differences

Average or Below Respect for Gender Differences	Strongly Agree/Agree	All Others	Total
White Female	0.0 (0)	100.0 (41)	41
Asian Female	13.3 (2)	86.7 (13)	15
Black Female	10.0 (1)	90.0 (9)	10
Hispanic Female	0.0 (0)	100.0 (12)	12
Total	3	75	78

Pearson chi-square test: Chi-square=6.795, p=.079, df=3
Cramer's V=.295

Table 4.16c: Men Make Better Team Members by Ethnicity for Females in Snobbish Atmospheres who Rate Professors Average or Below on Respect for Gender Differences

Average or Below Respect for Gender Differences	Strongly Agree/Agree	All Others	Total
White Female	0.0 (0)	97.6 (40)	41
Asian Female	20.0 (3)	80.0 (12)	15
Black Female	20.0 (2)	80.0 (8)	10
Hispanic Female	8.3 (1)	91.7 (11)	12
Total	6	72	78

Pearson chi-square test: Chi-square=8.757, p=.033, df=3

These results reported in Table 4.16a-c are based on tiny numbers of cases and cannot be considered the last word on this issue. However, I believe they will be corroborated in future research because they are intuitively appealing and, more important, they are consistent with findings from other tables that demonstrate that Asian women bring attitudes with them that make them vulnerable to doubts about their self worth and at heightened risk of dropping out without completing their degrees.

SUMMARY AND CONCLUSIONS

The relative absence of women from the top ranks of science and engineering is a huge problem, both from the standpoint of social justice and the health of the research personnel pool in the United States. For years, social scientists studying this problem have pointed to limited resources (see, for example Xie and Shauman, also Feldt, Mary Frank Fox), or women concentrated in community colleges and non-research oriented colleges and universities (see, for example, Valian). Cole and Singer, however, insisted that women's own reactions to the slights and disappointments, and generally their prior socialization, were key to understanding their lower publications productivity and relative absence from the ranks of top scientists and engineers.

In this chapter, I adduced considerable evidence that Cole and Singer were fundamentally correct. However, the graduate schools are, in my opinion, able to ameliorate the negative consequences of prior socialization if they make a concerted effort to address this issue.

The evidence from this chapter shows that, for example, Asian women many of them raised in societies where women were expected to be deferential to men such as their fathers and brothers and other male authority figures were much more likely in general than white women to believe men make better professors, or scientists, or team members. On the other hand, in welcoming environments, e.g., where the school atmosphere was regarded as friendly, where professors were rated above average on respect for gender differences, even Asian women were less likely to believe men make better professors, or better scientists, or better team members. In short, the self doubt, their lesser sense of self efficacy, and of competence, characteristic of women raised in environments where they were expected to be deferential to men, would be ameliorated to a considerable degree, perhaps even banished.

On the other hand, even among white women, a group where feelings that men were inherently better professors, or scientists, or team members was nearly unknown, I found that given conditions in the graduate school where the professors were average or below on respect for gender differences, where the atmosphere was snobbish, a significant minority responded that they believed men were better professors, or better scientists and engineers, or better team members.

The implications of these findings, even if they are not definitive, are nevertheless important. From a theoretical standpoint, they show the correctness of Cole and Singer's fundamental point that a lack of self efficacy resulting from prior socialization is causing proportionally more women and minorities than white men to leave science either in graduate school or after these women and minority scientists embark on their careers. And even if they do not leave science they shy from head to head competition with their more confident peers by avoiding work on the best problems in hot fields or they leave hot fields for less competitive specialties. Regardless, these less confident women and minority scientists do not have as much success, measured by years taken to reach full professorships, income, academic honors as the white men who dominate in the sciences. At the same time, the data suggest that the specific model Cole and Singer propose may be in need of revision. It may not be true that the accumulation of slights and disadvantages women and minorities have compared to white men is uniform. It is possible—indeed, it seems likely—that for native-born white women the main problems with self-efficacy begin in graduate school. For other women, including white women from societies where women are expected to be deferential to men, as well as for minorities, the accumulation of slights and disadvantages may begin earlier, perhaps as early as grade school. But it is probably simplistic to see this process of eroding of self-efficacy as somehow uniform. A discontinuous model may be more realistic.

The bottom line is that we need to recognize that graduate schools do much more in the way of molding future professionals than impart the latest and most advanced technical skills professionals will need to do quality work in their disciplines. Just as important, the graduate schools have a major impact on the attitudes of the students about their self worth and that of others of their gender or race. The graduate schools encourage friendships among students that will eventually turn into collaborations on cutting edge research projects and open doors to opportunities for professional advancement.

Graduate school professors and officials need to intervene to encourage self-efficacy. Many difficulties can be overcome in science by tenacity, tenacity that reflects the scientist's confidence that if he sticks to the thing, he will triumph. Getting across this point to budding scholars is necessary s if we are to address the shortages of scientists and engineers with advanced skills and assure American military security and economic prosperity in the information age.

Quite apart from changes graduate schools need to make in how they recruit, select, and prepare women and minorities for positions in STEM positions in academe, government, and business, they need to radically revise their recruitment and selection of foreign students. Foreign students in small numbers, perhaps up to ten percent of the total enrollment in a department, are highly desirable. They enrich the educational experience of all by bringing fresh perspectives and energy. Quite often they also pay full tuition—a welcome benefit to universities that are hard pressed for revenue. However, beyond a certain point—perhaps as little as ten percent or so—they are not desirable. First, unlike American students they are less likely to pressure their schools to maintain high standards. High standards are necessary or graduates will not be able to function in the demanding research and administrative positions of today's American economy. These incompetent graduates will not only cost their employers grievously; they will damage the reputation of the university from which they graduated, reducing opportunities for future graduates. Second, foreign students will likely return to their own countries to benefit their countries' businesses and possibly military establishments. Do we really want to train the leaders of the military industrial complex of China, India, or other potential adversaries at U.S. taxpayer expense?

Given these considerations, it is quite alarming to see that many of America's most prestigious research universities are keenly recruiting large numbers of foreign undergraduate and graduate students. The three universities with the most foreign students are: Columbia University in the City of New York has 26 percent of international enrollees in the country; Harvard University has 23 percent, and the University of Southern California has about 22 percent are (Mariano, 2013). In my opinion, the consequences of their policy to seek huge numbers of foreign students, driven primarily by a hunger for revenue, have dire long term consequences for the United States that are already visible in the economic prowess and military might of these formidable American competitors.

Chapter Five
Conclusions and Policy Implications

The United States faces a shortage of scientists and engineers with advanced credentials. Up to the present time, it has met the need for personnel with these credentials by a mix of recruiting abroad and graduating doctoral level scientists and engineers from its own universities. However, for a variety of reasons recruiting abroad may not yield enough doctoral level scientists and engineers in the future to meet the shortfall in numbers graduating from its own universities. These reasons include (1) policies making it more difficult for industry and academia to recruit abroad, some of which are "9/11" related; (2) less interest among white male students, who have been the traditional group from which the scientists and engineers have been recruited in studying for these degrees and lower fertility among white couples; (3) more opportunities for foreign trained scientists and engineers both in their own countries and elsewhere than the United States.

The solution to this shortfall in engineers and scientists, I maintain, is to step up recruitment from among women and minorities. In principle, academicians have accepted this as the preferred strategy. However, there are obstacles to effectuating this policy.

This country is still paying the price for an educational system that socialized girls and minorities to avoid science because it is too hard or too masculine or is all about things and has no interest in people. Thirty or so years ago, Sherman and Fennema (1977) documented that while even tenth grade and eleventh grade boys with mediocre mathematics aptitude took mathematics and science courses in high school, only girls in those grades with high aptitude in mathematics did the same. This is changing as is evident from the large numbers of women and minorities flocking to post secondary institutions and majoring in science and engineering. However, women who want to enter science or engineering are still a minority of the women with the necessary aptitude.

Part of the reason only a minority of women with the necessary aptitude is seeking to become scientists and engineers is what happens even now to promising girls. Perhaps they go on to college where they continue to shine in their mathematics and science courses. However, thanks to graduate school policies and practices that continue to the present and that push women and minorities out far more than they do white men, the bright futures of these promising young women scientists and engineers can be derailed.

Yet I am optimistic we could overcome the problem of too few women with aptitude in science seeking careers in that are if we were determined to devote the necessary resources to the task.

Money for education is available. The main obstacle is not lack of money but how we allocate it. Thanks to shortsighted American government policies we are not encouraging graduate education in the sciences and engineering. Money that could more productively be used for this purpose instead we spend on scholarships to send students to college where too often they select courses of study that are appealing to them but unlikely to lead to employable skills or to prepare them for rigorous graduate study in science and engineering. We fritter away money on questionable programs run by diploma mills and other "training organizations" that exist primarily to make a profit but graduate either few people or many with meaningless paper credentials.

While misallocation of money is the big problem, it is not the only obstacle we confront. I am convinced that the graduate school faculty themselves are skeptical about the great mass of students graduating from college, even those with majors in science. To my mind, the eagerness to recruit abroad seems as much to reflect this skepticism among senior faculty as it does the lack of funds for recruiting women and minorities, especially from other than elite institutions.

I have more faith in American students than I think many have. I think students can rise to the challenge if properly motivated. The graduate schools of our research universities can point to the good careers that await successful graduates. They can point to the employment advantages people with technical backgrounds enjoy in this economy. A college education may be better than a high school education. But recent experience shows that a technical education is better than a liberal arts degree. People with technical skills found jobs faster and received better salaries than people with liberal arts degrees during the recent "Great Recession."

We owe it to the American students to try to interest them in science and engineering careers. And we owe it to the American taxpayers as well. It is unreasonable and un-American to ask the taxpayer to fund a graduate department that educates mostly foreign students. True, these students pay more than in-state students in the public universities. But how much of the real cost of graduate school education is paid by foreign physics, chemistry, information sciences, and engineering students? Tuition in those programs is usually a fraction of the true cost. The income engines of universities are not the graduate science and engineering departments but the professional schools such as law, business, and maybe medicine. Even if all the students were foreign the graduate science and engineering departments might not pay their true cost.

Beyond being unfair to the taxpayer, to the extent these foreign students bring an ideology that is sexist, they also are creating a climate that is unwelcoming to women. This is unfair to all women—not just American women students but also women from other countries. If the cost of providing a program for women to overcome the sexist attitudes of the foreign students were factored in, would the tuition advantage be so much greater than providing spaces for more native-born students?

Before our cash strapped graduate schools rush pell-mell into recruiting abroad in countries such as China they should think through all the consequences. Are they prepared to take the necessary steps to make all foreign students welcome, not just men from a few Asian societies? They will almost certainly need to provide orientation to the students emphasizing that all people can do science and engineering. Men and women, Asian, white, Hispanic all have the ability to do the work.

Will the schools charge the true cost of the education for these students who in most cases because of legal restrictions will need to leave the country after they receive their degrees? Will the schools be as willing to help these students from abroad find good paying jobs after they graduate as they might for American students who will most likely seek positions in the United States? How will the schools handle the need for foreign students to have access to up to date equipment when much of it may be reserved for use by American citizens because of the sensitive nature of the technology?

If the graduate schools agree with me that they need to provide a better educational experience for their students, the schools will need to look at certain widespread practices that are counter- productive to producing more scientists and engineers. The destructive graduate school practices pushing promising young women out of science and engineering include teaching assistantships that, as I showed in CATSG, impose a large burden of undergraduate educational duties on graduate students, especially women. Now I have also shown they impose a heavier burden on African ancestry and Hispanic heritage students compared to white male students. Teaching assistantships need to be abolished, or, if that is impossible, to be radically reformed to prevent their being used to drive promising students out of science and engineering.

Sexism among graduate students in the S and E disciplines is not a serious problem overall since only a small minority of the students seem to have sexist attitudes. However, this is not ground for complacency. Schools that already have large foreign student populations (i.e., above ten percent of the student body in a given department) may have a sexism problem already of serious proportions. Schools that have weak policies on sexism, or where students see their professors as average or less than average in respect to gender differences, may also have a serious problem of sexism. And, if course, those that are seeking to balance their budgets by recruiting abroad may be facing a serious sexism problem down the line. For example, Ohio State University will need to monitor its student body carefully for sexist attitudes because it has opened offices in major cities in China—a country with a long history of female infanticide where despite official commitment to gender equality, women, especially in rural and inland areas, are expected to be deferential to men.

Sexism, as numerous writers have emphasized, not only both influences students to leave science and engineering. It also blights the careers of those women who bravely embark on careers in science and engineering despite discrimination. Sexism, when it is internalized, makes women students who have talent believe they are inferior, not good enough, and should pursue some other career. Even it is resisted, however, as seems to be the case with white women far more than African ancestry and Hispanic heritage women, sexism has adverse impacts on the women's careers as has been shown in numerous studies over several decades (see, for example, Bozeman and Corley, 2004; European Commission, 2003; Lorber, 1984; Rossiter, 1982). The women are less likely to find potential collaborators and are frozen out of discussions of new ideas that as Cole shrewdly observed are necessary to be productive in science. They are less likely to be evaluated positively than men with equal credentials (Fisher, 2007), more likely to see their blighted careers blamed on "their desire to have children" (European Commission, 2003; Williams and Ceci, 2012), or on attitudes that are not suited to science work at the frontiers as Pinker, 2005 might say. In short, as Cole pointed out, these women may be "in science" but they are not "of science".

I have previously urged that the graduate schools provide educational materials to students to acquaint them with the contributions of women and minorities to science and engineering. I renew that plea here. If the students understand that sexism and racism are destructive and have no place in the university they will have a more positive view of women and minorities and will (I hope and expect) carry those views with them when they complete their doctoral studies.

While I have specifically railed against teaching assistantships, recruiting foreign students, tolerance for sexism among students what I am really respect for the students as future scientists and engineers. If we continue to treat them as cheap labor while asking them to endorse the idea that women and minorities are worthy of respect I doubt we will be anywhere near as successful as we would if we made it clear we respect them and they should respect each other regardless of race, ethnicity, or gender.

Once we commit to a program of investing in the students, several steps that are specifically suggested in this study should occur as well. It makes sense, if we believe that students are future scientists and engineers that we invite speakers to talk about opportunities in the field. This conveys to the students that we want them to know about how they can apply their skills once they complete their degrees.

Informally urging students to start or continue research is another way in which we invest in the students. It came as a big surprise to me that in CATSG I found regional variations in answers to this question with the professors in the New York City/ suburban New Jersey region less likely to offer this encouragement than professors in the Research Triangle area or the Boulder/Denver area of Colorado. It is a small thing but it shows the professors actually care what the students do.

It would be naïve for me or anyone to believe this one slim volume will cause a sea change in American educational policy and practice. (Not since Harriet Beecher Stowe penned Uncle Tom's Cabin—whose author President Lincoln only half in jest remarked was the lady that started the Civil War—has a book had that kind of influence). It is not naïve to hope that it will cause thoughtful and concerned Americans to begin the informed dialogue that should lead to sound educational and economic policies and practices that will strengthen America down the line.

BIBLIOGRAPHY

Associated Press. "Out of college and out of work." *Albany Times-Union*, April 23, 2012.

———. "Bloomberg says tap world for talent." *Albany Times-Union*, May 23, 2012.

Astin, A.W. and H.S. Astin. *Undergraduate science education: The impact of different college environments on the educational pipeline in the sciences, Final Report to the National Science Foundation* (Grant Number SPA—8955365). Los Angeles: The Higher Education Research Institute, UCLA, 1992.

Bandura, Albert. "Self-efficacy theory in human agency." *American Psychologist* 37 (1977): 191-215.

Ball, John C. "The reliability ad validity of interview data obtained from 59 narcotic drug addicts." In *Stages of Social Research: Contemporary Perspectives*, edited by Dennis P. Forcese and Stephen Richer, 231-135. Englewood, NJ: Prentice-Hall, 1970.

Becker, Howard S. "Problems of inference and proof in participant observation." In *Stages of Social Research: Contemporary Perspectives*, edited by Dennis P. Forcese and Stephen Richer, 205-215. Englewood, NJ: Prentice-Hall, 1970.

Begley, Sharon. "He, once a she, offers own view on science spat." *Wall Street Journal*, July 13, 2006.

Bozeman, Barry and Sooho Lee. "New Study Shows Strong Association Between Academic Collaboration and Scientific Publishing Productivity," 2003. News release at http://gtresearchnews.gatech.edu/newsrelease/collaborate/htm.

Bozeman, Barry and Elizabeth Corley. "Scientists' Collaboration Strategies: Implications for Scientific and Technical Human Capital," *Research Policy* (2004).

Bush, Vannevar. *Science: the Endless Frontier*. Washington, D.C.: U.S. Government Printing Office, 1945.

Ceci, Stephen J. and Wendy M. Williams (editors). *Why Aren't More Women in Science?*. Washington, D.C.: American Psychological Association, 2007.

Cockburn, Cynthia. *Machinery of Dominance: Women, Men and Technical Know-How*. London: Pluto, 1985.

Cole, Jonathan R. "Women in Science" *The American Scientist* 69 (July-August 1981): 385-391.

Cole, Jonathan R. and Burton Singer. "A Theory of Limited Differences: Explaining the Productivity Puzzle in Science." In *The Outer Circle: Women in the Scientific Community*, edited by Harriet Zuckerman, Jonathan R. Cole, and John T. Bruer, 277-310. New Haven, CT: Yale University Press, 1992.

Cole, Jonathan R. and Harriet Zuckerman. "The Productivity Puzzle: Persistence and Change in Patterns of Publication of Men and Women Scientists." In *Advances in Motivation and Achievement*, edited by P. Maehr and M.W. Steinkamp, 217-258. Greenwich, CT: JAI Press, 1984.

Dean, Cornelia. "For some girls the problem with math is that they're good at it." *New York Times*, Feb. 1, 2005.

Economic Times. "Gen-next of immigrants in U.S. return home; India, China to gain from reverse brain drain." *Economic Times*. Accessed Oct 20, 2012. http://economictimes.indiatimes.com/news/nri/returning-to-India/gen-next-of-immigrants.

Ehrenreich, Barbara and Deirdre English. *Witches, Midwives, and Nurses: A History of Women Healers*. Old Westbury, New York: The Feminist Press, SUNY College at Old Westbury, 1973.

Epstein, Cynthia Fuchs. *Women's Place: Options and Limits in Professional Careers*. Berkeley, CA: University of California Press, 1970.

European Commission. *Creating Cultures of Success for Women Engineers Synthesis Report*. Prepared by Yvonne Pourrat of the CDEFI (France), 2003.

Etzkowitz, Henry, Carole Kemelgor and Brian Uzzi. *Athena Unbound: the Advancement of Women in Science and Technology*. Cambridge, England: Cambridge University Press, 2000.

Festinger, Leon. *A Theory of Cognitive Dissonance*. CA: Stanford University Press, 1962, (originally Rowe, Peterson & Co., 1957).

Fisher, Robert Leslie. *The Research Productivity of Scientists: How Gender, Organization Culture, and the Problem Choice Process Influence the Productivity of Scientists*. Lanham, MD: University Press of America, 2005.

———. *Making Science Fair: How Can We Achieve Equal Opportunity for Men and Women in Science*. Lanham MD: University Press of America, 2007.

———. *Crippled at the Starting Gate: The Graduate Schools Created and Perpetuate the Gender Gap in Science and Engineering: What Can We Do About It?*. Lanham MD: University Press of America, 2010.

_______. "Letter to the Editor," *American Scientist* (July-August, 2012) http://www.americanscientist.org/issues/pub/2012/4/scientists-and-motherhood.

Forcese, Dennis P. and Stephen Richer (editors). *Stages of Social Research: Contemporary Perspectives*. Englewood, NJ: Prentice-Hall, 1970.

Fox, L. H. and L. Brody and D. Tobin (editors). *Women and the Mathematical Mystique*. Baltimore, MD: Johns Hopkins University Press, 1980.

Fox, Mary Frank. "Gender, Environmental Milieu, and Productivity in Science." In *The Outer Circle: Women in the Scientific Community*, edited by Zuckerman, Cole, and Bruer, 188-204. New Haven and London: Yale University Press, 1992.

———. "Women and Scientific Careers." In *Handbook of Science and Technology Studies*, edited by Jasanoff, Sheila, Gerald E. Markle, James C. Petersen, and Trevor Pinch, 205-223. Thousand Oaks, CA: Sage Publications, 1995.

Ginther, Donna, Walter Schaffer, Joshua Schnell, Beth Masimore, Faye Liu, Laurel Haak, and Raymond Kington. "Race, Ethnicity and NIH Research Awards." *Science* 333 (2011): 1015-1019.

Haier, Richard J. "Brains, Bias, and Biology: Follow the Data." In *Why Aren't More Women in Science?*, edited by Ceci, Stephen J. and Wendy M. Williams, 113-120. Washington, D.C.: American Psychological Association, 2007.

Halpern, Diane F. "Science, Sex, and Good Sense: Why Women Are Underrepresented in Some Areas of Science and Math." In *Why Aren't More Women in Science?*, edited by Ceci, Stephen J. and Wendy M. Williams, 121-130. Washington, D.C.: American Psychological Association, 2007.

Hargens, L.L. *Patterns of Scientific Research: A Comparative Analysis of Research in Three Fields*. Washington, D.C.: American Sociological Association, 1975.

Herrnstein, Richard J. and Charles Murray. *The Bell Curve*. New York: The Free Press, 1994.

Higher Education Research Institute (HERI). *Survey of the American Freshman: National Norms 2000*. Los Angeles, CA: University of California-Los Angeles, 2001.

———. *Degrees of Success Bachelor's Degree Completion Rates among Initial STEM Majors*. Available at http://www.heri.ucla.edu/nih/downloads/2010%20-%20 Hurtado,%20Eagan,%20Chang%20-%20Degrees%20of%20Success.pdf.

Hsu, Steve. "Letter to the Editor." *American Scientist*, July/August, 2012.

Hyde, Janet S., Sarah M. Lindberg, Marcia C. Linn, Amy B. Ellis, and Caroline C. Williams. "Gender Similarities Characterize Math Performance." 321, no. 5888 Science (July 25, 2008): 494-495.

Jasanoff, Sheila, Gerald E. Markle, James C. Petersen, and Trevor Pinch (editors). *Handbook of Science and Technology Studies*. Thousand Oaks, CA: Sage Publications, 1995.

Kaminski, Deborah and Cheryl Geisler. "Survival Analysis of Faculty Retention in Science and Engineering by Gender." 335, no. 6070 *Science* February 2012): 864-866.

Korbe, Tina. "U.S. faces a reverse brain drain, as highly educated immigrants head home." April 17, 2012. http://hotair.com/archives/2012/04/17/u-s-faces-a-reverse-brain-drain-as-highly-educated-immigrants-head-home/.

Korn, Melissa. "Chinese Applicants Flood U.S. Graduate Schools." *The Wall Street Journal*, April 3, 2012.

Leslie, Larry L. and Gregory T. McClure and Ronald L. Oaxaca. "Women and Minorities in Science and Engineering: A Life Sequence Analysis." *Journal of Higher Education* 69, no.3 (May/June, 1998): 239-276.

Langberg, Kamma. "Gender Gap and Pipeline Metaphor in the Public Research Sector," Organization of Economic Cooperation and Development: Workshop on Women in Science, 2005.

Lippa, R. "Gender related individual differences and the structure of vocational interests of the people-things dimension." *Journal of Personality and Social Psychology* 74 (1998): 996-1009.

Little, Daniel. "False consciousness: A brief explanation of Marx's conception of false consciousness; some of the ways in which later Marxist thinkers have used the concept. http://www-personal.umd.umich.edu/~delittle/iess%20false%20conscious ness %20V2.htm.

Lorber, Judith. *Women Physicians: Careers, Status, and Power*. New York and London: Tavistock Publications, 1984.

Mariano, Jules. "Most Popular Universities in the U.S. for International Students." http://ezinearticles.com/?Most-Popular-Universities-in-the-US-for-International-Students&id=7468021.

National Action Council for Minorities in Engineering. *2011 NACME Data Book: A Comprehensive Analysis of the "New" American Dilemma, 2011*. Available at http://www.nacme.org/user/docs/NACMEDatabookReprintFinal2Post.pdf.

National Center for Educational Statistics, "Fast Facts", U.S. Department of Education 2011. http://nces.edgov/fastfacts/display.asp?id=72.

Neuliep, James (editor). *Replication Research in the Behavioral Sciences*. Newbury Park, CA: Sage Publications, 1991.

New York Times. "New Light on Origins of Ashkenazi in Europe." *New York Times*, January 14, 2006.

Osen, L. M. *Women in Mathematics*. Cambridge, MA: MIT Press, 1974.

Payne, Stanley L. *The Art of Asking Questions*. Princeton NJ: Princeton University Press, 1980 (originally 1951).

Pinker, Steven and Elizabeth S. Spelke. *The Science of Gender and Science: Pinker vs. Spelke*. May 10, 2005. www.edge.org/3rd_culture/debate05/debate05_index.html.

Rosenthal, Robert. "Replication in Behavioral Research." In *Replication Research in the Behavioral Sciences*, edited by James W. Nieuliep, 1-30. Newbury Park, CA: Sage Publications, 1991.

Rossiter, Margaret W. *Women Scientists in America: Struggles and Strategies to 1940*. Baltimore, MD: Johns Hopkins University Press, 1982.

Sherif, Muzafer, O.J. Harvey, B. Jack White, William R. Hood, and Carolyn W. Sherif. *The Robbers Cave Experiment: Intergroup Conflict and Cooperation*. Middletown, CT: Wesleyan University Press, 1988 (originally Norman, OK: University of Oklahoma Press, 1961).

Sherman, J. and E. Fennema. "The Study of Mathematics by High School Girls and Boys: Related Variables," in *American Educational Research Journal* 14 (1977): 159-168.

Sinostand.com. "The Gaokao Highway to Hell." http://sinostand.com/2013/02/03/the-gaokao-highway-to-hell/

Tang, Joyce. *Doing Engineering: the Career Attainment and Mobility of Caucasian, Black, and Asian-American Engineers*. Lanham, MD: Rowan and Littlefield Publishers, 2000.

Valian, Virginia. *Why So Slow: The Advancement of Women*. Cambridge MA: MIT Press, 1998.

Wajcman, Judy. "Feminist Theories of Technology." In *Handbook of Science and Technology Studies*, edited by Jasanoff, Sheila, Gerald E. Markle, James C. Petersen, and Trevor Pinch, 189-204. Thousand Oaks, CA: Sage Publications, 1995.

Ware, Norma C., Nicole A. Steckler, and Jane Leserman. "Who Chooses a Science Major?" in *Journal of Higher Education* 56, no. 1 (1985): 74-84.

Warren, Wini. *Black Women Scientists in the United States*. Bloomington, IN: Indiana University Press, 1999.

Waters, Mary C. *Black Identities: West Indian Dreams and American Realities*. Boston: Harvard University Press, 2001(originally1999).

"Why International Recruitment Matters." *American International Education Foundation*. Accessed May 1, 2013. http://aief-usa.org/resources/recruitment/recruitment_why_page2.htm.

Williams, Wendy M. and Stephen J. Ceci. "When Scientists Choose Motherhood," American Scientist (March-April, 2012) 138-145.

Wilson, Robin. "Gender Equity on Science Faculties Might Have to Wait a Century, Study Finds," Chronicle of Higher Education, February 16, 2012.

Xie, Yu and Kimberlee A. Shauman. "Sex differences in Research Productivity: New Evidence about an Old Puzzle," American Sociological Review 63 (1998) 847-870.

———. *Women in Science: Career Processes and Outcomes*. Cambridge, MA: Harvard University Press, 2003.

Zuckerman, Harriet, Jonathan R. Cole and John T. Bruer (editors). *The Outer Circle: Women in the Scientific Community*. New Haven and London: Yale University Press, 1992.

INDEX

Page numbers followed by the letter "t" denote tables.

E

F

G

J

K

L

M

N

O

P

R

S

W

X

www.ingramcontent.com/pod-product-compliance
Ingram Content Group UK Ltd.
Pitfield, Milton Keynes, MK11 3LW, UK
UKHW012041150926
475560UK00001B/131

9 780761 862581